S0-BIS-497

From Disaster Response to Risk Management

Advances in Natural and Technological Hazards Research

VOLUME 22

EDITORIAL BOARD

Wang Ang-Sheng, *Chinese Academy of Sciences, Institute of Atmospheric Physics, Beijing, P.R. China*

Gerhard Berz, *Münchener Rückversicherungs-Gesellschaft, München, Germany*

Oscar González-Ferrán, *Departamento de Geologia y Geofisica, Facultad de Ciencias Fisicas y Mathematicas, Universidad De Chile, Santiago, Chile*

Cinna Lomnitz, *National University of Mexico, Instituto de Geofisica, Mexico, D.F. Mexico*

Tad S. Murty, *Baird & Associates, Ottawa, Ontario, Canada*

Alvin H. Mushkatel, *Office of Hazards Studies, Center for Public Affairs, Arizona State University, Tempe, AZ, USA*

Joanne M. Nigg, *Disaster Research Center, University of Delaware, Newark, DE, USA*

Alexei V. Nikolaev, *Institute of Physics of the Earth, Russian Academy of Sciences, Moscow, Russia*

Paul M. Thompson, *Flood Hazard Research Center, Middlesex University, Enfield, UK*

Donald A. Wilhite, *International Drought Information Center, University of Nebraska, Lincoln, NE, USA*

The titles published in this series are listed at the end of this volume.

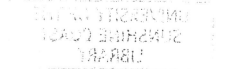

TABLE OF CONTENTS

FOREWORD

In 1992, Australia's Commonwealth and State governments announced the introduction of the National Drought Policy. In an important departure from earlier responses to drought, the new policy adopted a risk management approach which received broad support from Australia's major political parties and from important members of the policy community. More than a decade later, however, media and public debate about drought does not always reflect this policy position.

In 2002, Linda Botterill and her colleague Melanie Fisher invited Don Wilhite to contribute a chapter to a book on Australia's National Drought Policy which they were preparing for an Australian audience. Subsequently published as *Beyond Drought: People, Policies and Perspectives* by CSIRO Publishing, the collection was intended 'to place drought on the public agenda as a topic of considerable importance to all Australians' (Botterill and Fisher 2003, p ix). The book was targeted at the interested lay public in an effort to redress the gap between the official policy position and public perceptions of drought. Don Wilhite was invited to contribute a chapter placing Australia's experience in an international context. During a visit to Australia to discuss the scope and content of the book, Don suggested that the themes being discussed would be of considerable interest to an international audience and he proposed that a more academically focused collection along similar lines to the Australian publication would provide a useful addition to the international literature on drought preparedness and response.

The present collection is the result of that suggestion. The majority of the material in this book is new, indeed four of the authors included herein did not participate in the earlier project and of those who did, most have contributed entirely new work. A handful of the chapters contain material published in the Australian book, but with the exception of one chapter the material has been reworked for a broader audience.

Australia has been something of a trail-blazer in the development and implementation of a national drought policy. It is hoped that this collection will provide others moving in a similar direction with the benefit of its experience by highlighting the successes and challenges of a move from disaster to risk management in responding to drought.

Linda Courtenay Botterill
Donald A Wilhite
April 2004

CONTRIBUTORS

Linda Courtenay Botterill is a Postdoctoral Fellow in the National Europe Centre at the Australian National University. She has extensive experience in public policy having worked in the Australian Public Service, as a ministerial adviser and as a policy officer in two industry associations before undertaking her PhD in political science at the ANU. In 1993 and 1994, she advised the Australian Minister for Primary Industries and Energy on drought and rural adjustment. Her research interest is agricultural policy in Australia and the European Union and she teaches in the field of Australian public policy. She is co-editor of *Beyond Drought: People, Policy and Perspectives* (2003 CSIRO Publishing).

Peter Cox was awarded a PhD in Technological Economics from Stirling University in Scotland. He worked in several developing countries in Africa and SE Asia. He spent eight years with CSIRO (at Narrabri with the Division of Plant Industry, and Toowoomba with the Division of Tropical Crops and Pastures) after moving to Australia from Papua New Guinea in 1988. More recently, he worked with the International Crops Research Institute for the Semi-arid Tropics (ICRISAT) in India and the International Rice Research Institute (IRRI) in Cambodia as part of their social science programmes. His last position was as Regional Technical Adviser for Agriculture and Natural Research Management for South East Asia with Catholic Relief Services, a US-based NGO. Peter Cox died from cancer late in 2003 during preparation of the chapter for this book. He is missed more than he, with his characteristic modesty, would have imagined.

Peter Hayman is Principal Scientist, climate applications working for the South Australian Research and Development Institute, based in Adelaide. From 1999 to 2004 he led the NSW Agriculture agro climatology unit. He is a member of the World Meteorological Organisation expert team on end user liaison and the expert team on weather climate and farming. After completing a masters degree in crop physiology, Peter Hayman worked as an extension officer before completing a PhD in agro climatic risk management at the University of Western Sydney. He has received research grants to work with grain farmers on their management of climate risk in the north-eastern and southern grain belt and with wool producers in the NSW rangelands.

Greg Hertzler is a Senior Lecturer in the School of Agricultural and Resource Economics at the University of Western Australia and was Deputy Chair of the recent West Australian Task Force into Multi Peril Crop Insurance. His research interests are in agriculture and natural resources with particular emphasis on bio economic modelling and on managing agricultural and natural systems under risk.

Janette Lindesay has a PhD in Climatology from the University of the Witwatersrand in South Africa, where she led the Climatology Research Group in research into southern African climate variability and drought in the early 1990s. In 1993 she took up an appointment as Senior Lecturer in Climatology at the ANU, where she is engaged in research on low-frequency variability in the El Niño Southern Oscillation and its impacts, and on changing rainfall seasonality in Australia. She has co-authored a number of papers and an authoritative text on ENSO and climate variability, and has co-authored or edited three other books. Her other research interests include climate change, climate impacts and climatological aspects of wildfire in the tropics and subtropics.

Bruce O'Meagher is a former senior public servant. He worked in several national government agencies, including the Treasury and the agriculture and industry departments. He was heavily involved in the development of the Australian government's response to the 1992-95 drought.

Deborah Bird Rose is Senior Fellow in the Centre for Research and Environmental Studies, Institute of Advanced Studies, at The Australian National University. She is the author of *Country of the Heart: An Indigenous Australian Homeland* (2002 Aboriginal Studies Press), *Nourishing Terrains, Australian Aboriginal views of Landscape and Wilderness*, *Dingo Makes Us Human* (winner of the 1992/3 Stanner Prize), and *Hidden Histories* (winner of the 1991 Jessie Litchfield Award). She has worked with Aboriginal people in their claims to land, and is collaborating with the New South Wales National Parks and Wildlife on their 'Totemic Landscapes' project. Her work in both scholarly and practical arenas is focused on social and ecological justice. Her most recent book is in press with UNSW Press: *Reports from a Wild Country: Ethics for Decolonisation*.

Mark Stafford Smith has worked at CSIRO's Centre for Arid Zone Research in Alice Springs for two decades, with an emphasis on management responses to climatic variability in rangelands grazing enterprises for most of that time. He has had several encounters with the design, implementation and implications of drought, from input to the National Drought Policy, through surveys and modelling of pastoralists' responses to drought in different rangeland regions, and involvement in regional adjustment committees, to on-farm analyses of the effects of taxation and other policy instruments on pastoralist decision making. He is currently CEO of the Desert Knowledge Cooperative Research Centre, based in Alice Springs and networking colleagues across the continent towards better regional outcomes for all people living in outback Australia.

Daniela Stehlik returned to Western Australia from Queensland in December 2003 to take up the foundation Chair in Stronger Communities in the Division of Humanities at Curtin University of Technology. She has also been appointed as Director of the Alcoa Research Centre for Stronger Communities, a unique partnership between industry/community/university. Her research interests focus on the intersections of community resiliency, human service practice and social cohesion in regional/rural Australia. Her specific interests are in ageing, disability, gender, power and community development. She has published widely in Australia and internationally and is currently on the editorial board of *Rural Society*.

Ian Ward is a Reader in Politics at the University of Queensland. His research interests lie broadly in the area of media and politics.

David White is the Director of ASIT Consulting based in Long Beach, NSW. From 1967 until 1988 he worked for the Victorian Department of Agriculture, supervising the development and use of models to analyse livestock production systems. He then moved to Canberra where, as a Senior Principal Research Scientist in the Bureau of Resource (now Rural) Sciences, he assisted in the development and implementation of the National Drought Policy. From 1994 until late 1996 he was the principal scientific adviser to the Rural Adjustment Scheme Advisory Council (RASAC), primarily with respect to drought monitoring and assessment and Drought Exceptional Circumstances. His consulting activities over the past seven years have targeted a wide range of environmental and agricultural issues pertaining to climate variability research, climate change and increasing water use efficiency in the Murray-Darling Basin.

Donald Wilhite is Director of the National Drought Mitigation Center and the International Drought Information Center and Professor, School of Natural Resources, University of Nebraska-Lincoln. Dr. Wilhite's research and outreach activities focus on issues of drought monitoring, planning, and mitigation. He has collaborated with numerous countries and regional and international organisations on drought policy and planning issues. He has authored or co-authored more than 100 journal articles, monographs, book chapters, and technical reports and is editor of several books, including *Drought: A Global Assessment*, published in 2000 by Routledge Publishers as part of a 7-volume series on natural hazards and disasters.

ABBREVIATIONS

ABARE	Australian Bureau of Agricultural and Resource Economics
AUSLIG	Australian Surveying and Land Information Group
CQ	Central Queensland
CSIRO	Commonwealth Scientific and Industrial Research Organisation
DPRTF	Drought Policy Review Task Force
EC	Exceptional circumstances
ENSO	El Niño-Southern Oscillation
FMDs	Farm Management Deposits
GDP	Gross Domestic Product
GDPN	Global Drought Preparedness Network
IAC	Industries Assistance Commission
IPCC	International Panel on Climate Change
MEI	Multivariate ENSO Index
NDMC	National Drought Mitigation Center (University of Nebraska, Lincoln)
NDP	National Drought Policy
NDRA	Natural Disaster Relief Arrangements
NGOs	Non-governmental organisations
NSW	(the state of) New South Wales
OECD	Organisation for Economic Cooperation and Development
RIRDC	Rural Industries Research and Development Corporation
SOI	Southern Oscillation Index
TEK	Traditional Ecological Knowledge
UNCCD	United Nations Convention to Combat Desertification
UNESCO	United Nations Educational, Scientific and Cultural Organisation
WA	(the state of) Western Australia

GLOSSARY

Australian Labor Party	Social democratic political party
battlers	Australians with relatively low incomes facing economic hardship or "doing it tough" [a related term]
bush	Traditional term describing rural and regional Australia
Commonwealth government	(see federal government)
federal government	Australia's national government. The dominant tier of government within a federal system, raising most of the taxes paid by Australians
Liberal Party	Conservative party—traditionally governs in coalition with the National Party
mateship	Traditional term for friendship but carrying additional egalitarian connotations
National Farmers' Federation	Major rural lobby group comprising state farm organisations and commodity councils and representing some 120,000 farm enterprises through 29 affiliated organisations
National Party	Formerly the Country Party—founded to provide a voice for rural Australia
NSW	New South Wales, Australia's most populous state, which has Sydney as its capital
Queensland	Australia's second largest and most decentralised state
rort (v)	Australian colloquialism for trick, dishonest practice

INTRODUCTION

LINDA COURTENAY BOTTERILL
National Europe Centre, 1 Liversidge Street (#67C), Australian National University, ACT 0200, Australia

DONALD A WILHITE
National Drought Mitigation Center and International Drought Information Center, University of Nebraska–Lincoln, 239 LW Chase Hall, Lincoln, Nebraska 68583, USA

Australia's attempt at a national drought policy in the early 1990s and its experiences with this policy over the past decade have intrigued the international scientific and policy communities. Few nations have made much progress on a national policy but it is now being widely discussed by many countries and promoted by United Nations agencies, international development organizations, development banks, and others. For example, under the United Nations Convention to Combat Desertification (UNCCD), countries are encouraged to develop national action programs to combat the effects of desertification and drought. There are also many other national, regional, and global initiatives to promote the need for greater levels of drought preparedness and to formulate national drought policies. The experiences of Australia represent valuable lessons to many countries, developed and developing alike, on the opportunities and challenges of a national drought policy and preparedness strategy. Documenting the policy development process and the lessons learned at each step in the process will benefit all nations that choose to follow this course of action.

For example, the United States has drawn on the experiences of Australia in recent attempts to move towards a national drought policy. Until recently, much of the progress in drought preparedness in the United States has been at the state level as the need for and benefits of drought planning have become more apparent. With the increase in the number of states with drought plans from 3 in 1982 to 36 in 2004, the existence of actual drought plans for states to follow in the plan development process has certainly stimulated this planning trend. Drought plans enable states to visualise how others have applied planning methodologies to meet their specific drought management needs. Likewise, Australian experiences with drought policy have been beneficial to the drought policy debate in the United States. Legislation is now pending before the US Congress on a national drought policy action plan.

However, the implementation of Australia's National Drought Policy has not been without its problems. Tensions within the policy between conflicting objectives have led to ongoing changes to the various components of the policy and, although the rhetoric of self-reliance and risk management is accepted within the rural policy community, the Australian media, the broader community and many farmers appear to retain a view of drought as a disaster or an Act of God. This volume brings together a range of perspectives on Australia's experience since the announcement of a National

L.C. Botterill and D.A. Wilhite (eds.), From Disaster Response to Risk Management, 1–4.
© 2005 *Springer. Printed in the Netherlands.*

Drought Policy in 1992 in order to illustrate the challenges that have faced policy makers in implementing a policy approach which takes a more realistic view of the highly variable nature of Australia's climate.

We begin the collection with Mark Stafford Smith's chapter, which gives us an overview of the Australian environment, highlighting some of the features which make this continent unique. Janette Lindesay's chapter builds on this, addressing the inherent variability of climate in the sub-tropical regions of the Earth, with an emphasis on Australia's climate. A range of drought definitions is considered, since different components of human and natural systems respond to a rainfall deficit in different ways. This often complicates the assessment of the beginning and end of drought conditions. An overview of the drought history of the continent highlights the impacts on Australia of the most severe droughts of the last century. The importance of using an understanding of past climatic variability, and drought in particular, to inform both current decision making and policy formulation and planning for the future is highlighted in a discussion of scenarios for possible future climate change in Australia.

Australian's indigenous people have lived with the uncertainty of the continent's climate for a staggering 40,000 years or more. In her chapter, Deborah Bird Rose describes the centrality of water to aboriginal cultural practice and explores indigenous knowledge systems. She examines several aspects of knowledge and practice that are particularly illuminating of people's achievements in understanding and living with a great deal of uncertainty.

In the late eighteenth century, Europeans arrived in Australia and brought with them an understanding of climate that regarded drought as an aberration and predictability as the norm. Rather than adopt the survival practices of the indigenous people, they introduced the agricultural technologies and approaches with which they were familiar. Against this cultural background, and for more than two centuries, droughts were regarded as natural disasters against which farmers and the broader community did battle. This changed in 1989 with an important decision to remove drought from events covered by national natural disaster relief arrangements. Linda Botterill's chapter outlines the policy developments of the late twentieth century which resulted in the 1992 announcement of a national drought policy based on principles of self-reliance and risk management.

As Daniela Stehlik illustrates in her chapter, this paradigmatic shift, from disaster response to risk management, has not been universally accepted. Based on a major research project undertaken during the drought of the 1990s, she describes the impact of the new policy approach on farm families. Drawing on the voices of those most directly affected, Stehlik presents their responses to the shift from drought as 'disaster' to drought as 'managed risk' and frames them within a broader discussion of the impact of social policy in response to drought on the lived experiences of Australian families and communities.

During the last two major droughts in Australia, the 1990s and 2001-03, major media organisations have organised public 'Farm Hand' appeals to raise funds for drought-affected farm families. These appeals and the response to them by the general public suggest that the policy shift has gone largely unnoticed by the media and the general public. Ian Ward's chapter examines news reporting of drought in Australia. He argues that droughts pose a particular set of problems for journalists. Drought is difficult to define, drawn-out, and doesn't generate the graphic images that natural disasters such as storms, floods and bush fires will. News coverage often involves familiar images of dusty paddocks, drying dams and dying stock, and focuses attention on the plight of farm families battling against the odds. The formulaic manner in which metropolitan news media report and 'frame' droughts creates a public awareness of, and sympathy for, struggling farm families. Insofar as news coverage drives the policy agenda, when drought disappears from the front pages there is little opportunity to drive policy makers to consider alternative, longer-term policies aimed at encouraging the adoption of farm practices more suited to Australian climatic fluctuations.

Although focused on risk management, Australia's National Drought Policy retained an element of the previous 'disaster' approach with the creation of the concept of 'exceptional circumstances' as a trigger for enhanced government assistance. This feature has been at the heart of the policy's implementation problems. David White, Linda Botterill and Bruce O'Meagher address the difficulties that have been encountered in arriving at an operational definition of 'exceptional circumstances', highlighting the fact that, although the science is promising in its capacity to produce a trigger point, the politics of drought relief in a liberal democracy limit the capacity of policy makers to introduce a defensible, objective trigger for exceptional drought declarations.

The design and modification of effective policy instruments to improve the management of drought risk in Australia depends in part on understanding how farmers perceive and manage the risks associated with periodic drought. In their chapter, Peter Hayman and Peter Cox examine these risks, how farmers perceive them, how scientists try to quantify them, how quantification can be used to improve risk management, and some of the pitfalls in relying too much on 'scientific' models for risk assessment. Mismatches are noted between different perspectives of drought as risk between farmers, scientists and policy economists. These mismatches suggest that the issue of drought is still incompletely specified and that its complexity will always allow multiple interpretations.

Adopting a risk management approach to drought implies that farmers will have access to tools to manage their risk. Greg Hertzler's chapter reviews the prospects for crop insurance, weather derivatives and yield index contracts for insuring against drought in Australia. Over the previous century, multiple-peril crop insurance schemes have failed many times in many countries. Most surviving schemes are subsidised by governments. Australia, too, has its share of failures, and three separate reports have recommended against government support for multiple-peril crop insurance. An alternative to crop insurance is rainfall insurance, first proposed by Australian researchers in the 1980s. This idea has been transformed into weather derivatives. Yield index contracts have been proposed as an intermediate between crop insurance and weather derivatives. Hertzler's chapter explores the advantages and disadvantages of these different instruments and the role they might play in the implementation of the National Drought Policy.

Following the discussions on the social impacts of drought and the difficulties associated with implementing a risk management approach, Bruce O'Meagher returns to first principles and, taking an economics perspective, discusses the rationales for government intervention in the event of drought. It is accepted that drought has economic consequences and governments in most countries have intervened to manage those consequences. However, not all of these economic consequences are adverse and much of the intervention that does occur is counterproductive. He measures the National Drought Policy against its objectives and suggests some improvements that could enhance the efficiency and effectiveness of the policy response. He also highlights the conflicting nature of the objectives the policy sets out to achieve.

Don Wilhite's chapter puts Australia's experience in a broader international context, illustrating that, along with the US, Australia is something of a world leader in the development and implementation of drought policy based on preparedness and risk management. The chapter describes the drought policy experience in the US and sub-Saharan Africa and emphasises the importance of regional approaches. It points to the need for international co-operation and information-sharing on drought management—including through a Global Drought Preparedness Network. This book hopes to contribute to that process of information-sharing by bringing to the attention of a wider audience the challenges which have faced Australian policy makers and the lessons that can be learnt from more than a decade of experience of implementing a national drought policy.

We conclude the book with an overview of the broad themes raised by our contributors.

CHAPTER 1: LIVING IN THE AUSTRALIAN ENVIRONMENT

MARK STAFFORD SMITH*
Desert Knowledge Cooperative Research Centre. PO Box 2111, Alice Springs, NT 0871, Australia

1. Introduction

For 100 years after the Europeans arrived in Australia, their painters portrayed a landscape which harked back to the soft lights of Europe, their poets wrote wistfully of their respective mother countries, and their music was folk songs from rural Britain. In the context of this culture, rooted in Europe, men of the land (it was mainly men) carried out farming as if it was merely a matter of applying a fine work ethic to subdue the country into a reliable European image.

Around the end of the 19th-century, artists began to portray a harsher countryside, poets to extol a more local larrikin approach to life, and composers began to incorporate images of the bush peculiar to Australia like the kookaburra into their music. In 1894, the first major scientific expedition to central Australia, the Horn Expedition, reflected these changing views (Spencer 1994). Among its participants there were still those seeing a landscape with European eyes; but others, most notably Baldwin Spencer, had begun to see how the country varied spatially as they moved through it on their camels, and through time as he visited repeatedly over the following years. Spencer met pastoralists who were beginning to understand and take advantage of this new land. But his descriptions still make it clear that most inhabitants regarded the ups and downs of climate as being an unfair imposition from on high, rather than a normal feature of the environment, to be managed and celebrated.

This chapter briefly outlines the special features of the Australian physical and social environment which affect the way we manage our non-urban areas. The outback, which has had a deeply symbolic place in the way Australians view their country, today mainly refers to the arid and semiarid interior. What has been viewed as outback has changed over the years—at one time everywhere across the Blue Mountains was included, but the boundary between outback and inside country slowly flowed towards the interior as settlement proceeded. However, issues of drought management affect all non-urban areas and many of the issues which are writ large in the arid zone still affect other regions in a slightly more subdued fashion.

*This chapter was first published in Beyond Drought: People, Policy and Perspectives; Linda Courtenay Botterill and Melanie Fisher (eds). CSIRO PUBLISHING, Melbourne 2003. Reproduced by permission of the Publisher.

2. The Biophysical Environment

Key features of the Australian biophysical environment have been summarised many times (see for example AUSLIG 1992; Beadle 1981; Friedel *et al* 1990; Groves 1994; NLWRA 2002; Stafford Smith 1994a). It is easy to generalise across the whole continent, and such statements must always be tempered with the real diversity of more local conditions. This will become important when we come to consider the scale at which issues such as drought should be managed. For example, we inevitably focus on the nature of climate in Australia, but as Figure 1 shows, there is an immense diversity of climatic patterns represented in the continent, from the relatively reliable monsoonal systems in the North through the incredibly uncertain arid centre to the somewhat more reliable temperate southern systems. We will return to this issue.

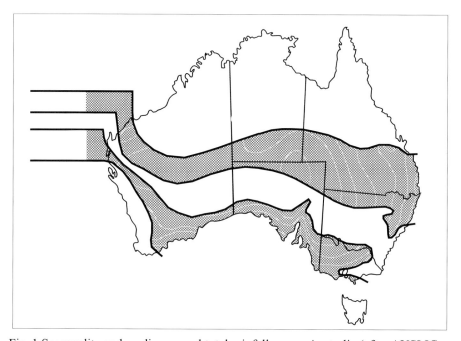

Fig. 1 Seasonality and median annual total rainfall across Australia (after AUSLIG 1992)

At the base of it all, this is an ancient continent, worn into low relief by millions of years of exposure to the changing atmosphere, and concentrated into salt lenses and silcretes by an equal period of leaching. Coupled with the resulting low productivity, at least for the last few tens of thousands of years, the continent has been located in a particularly variable part of the earth's climate system. On top of the normal annual variability found in all semi-arid areas in subtropical to temperate zones of the planet, Australia experiences additional multi-annual variability drivers such as El Niño (Nicholls and Wong 1990). We are also increasingly recognising that there is an inter-decadal timescale of variability that may be specifically modulated by the 'Inter-decadal Pacific Oscillation' but is affected by other long cycle events in the world's oceans; and of course we may now be facing directional climate trends driven by global climate change. Significant rainfall events are immensely more diverse for an Australian site than for a US site in a comparable mean annual and seasonal rainfall regime.

These features alone have major implications for the Australian biota (see for example Barker and Greenslade 1982; Dodson and Westoby 1985; Saunders *et al* 1990; Stafford Smith and Morton 1990). Plants living in the US (or Mediterranean) environments can be reasonably confident that they will receive another rainfall of a given size with a consistent return time, usually less than a year. By comparison, the Australian plant (speaking teleologically!) has no idea when its next drink is going to fall. In consequence, stem succulence is a highly successful strategy in plants in Central America, whilst species with these characteristics are almost entirely absent from Australia (Stafford Smith and Morton 1990). This is not a surprising response to the fact that, whilst you can store a water supply in your stem with confidence of replacement within a year in America, this would be (or has been) a suicidal recipe in Australia with its uncertain climate. As another example, in a regular climate where you know there will be a time of year through which it is hard to persist, it makes sense to be deciduous—to drop your leaves and reduce your costs of obtaining water at times when it is in short supply. But if you live in Australia and you don't know when these times are going to occur nor when they are going to end, your leaves become a more precious resource. This effect is exaggerated still further if you live on poor soils where nutrients are at a premium, and you don't want to be re-growing your leaves from scratch on a regular basis. For these reasons and others, Australian plants tend not to be seasonally deciduous, though some exhibit drought-deciduousness, having a strategy for slowly dropping leaves as conditions get drier.

Such ecological linkages may seem quite esoteric, but they have major implications for native or domestic animals that live on these resources, and for sustainable land management. Our native animals have come to their own accommodation with these underlying ecological drivers. The classic example often cited is the kangaroo, with its suppressed embryo developments, such that if one joey is aborted there is a new foetus instantly ready to take its place when conditions become favourable. But there is a multitude of comparable relationships that are more subtle but systemically far more significant, for example driving the ecosystem services provided to agriculture by ants and termites. Again, these have been discussed in a variety of places, although the way in which their implications flow through to agricultural management is still a matter for considerable debate.

Two immediate examples related to management are fire and grazing. Because plants on poor soils are relatively more constrained by nutrients and by the production of carbohydrate, many Australian ecosystems produce plentiful and long-lived fuel, and are consequently subject to fire. There are many other reasons why fire is an important force in the Australian ecosystems, and why many Australian species may have evolved to 'use' fire, but the result is that management must cope with and take advantage of the effects of fire. Our slow development of understanding about the intimate relationships between fire and vegetation composition in our landscapes has led to a variety of problems, including woody weed invasion in rangelands and the loss of resilience in the management of our forest ecosystems, which contribute to or contrast with the issues related to drought management.

Grazing also interacts with the features of Australian ecosystems, not surprisingly. There are a variety of different pasture types within grazed Australia; a significant proportion of these are dominated by native plants that can cope quite well with intermittent catastrophic losses of foliage but very poorly with chronic defoliation such as that caused by continuous grazing. Thus we have seen widespread loss of palatable perennial plants in Australia's grazed lands. At the same time, while early settlers soon found and exploited those patches of country which had richer soils and which carried plants that were less constrained by nutrients than on the general landscape, we are still learning how to manage this complex mosaic of environments. The problem is that the overall general low productivity of many of our farmed systems means that individual management units (as such as paddocks) must be large, and consequently end up including a diversity of landscape types. This makes them much harder to manage than small fields which can segregate chunks of the landscape with different characteristics and then manage them differentially (Stafford Smith 1994a).

Agriculture and forestry too must deal with these underlying characteristics of our ecosystems. Forests contain a mosaic of types of trees in much the same way that rangeland environments contain mosaics of different perennial grasses. Richer patches of soil support trees and grasses which tend to be more productive, and are associated with a particular suite of native animals that require food of higher quality. Just as cows may compete for food by selectively grazing the better quality forage, so selective logging for faster growing types of tree may often compete with a certain suite of native animals for this resource. In both cases, those ecosystems which depend on the richer soils tend to be selectively used, or even over-used, and their dependent ecosystems thereby put at risk. These richer pockets in a generally poorer landscape are also the areas that are most completely cleared for intensive agriculture, so that the same issue arises in a different form in agricultural landscapes.

Why do all these issues matter for drought management? The generality behind them is that the intensification of ecosystem use, whether from native animals to pastoral grazing, or from grazing to intensive agriculture, is a process of gradually replacing the ecosystems' internal buffering dynamics by external management and subsidy. For example, in moving from kangaroos to cattle, we move from a system in which kangaroo populations boom in good years and move or die back in poor years, to a system in which we constrain grazing animals from moving so freely across the landscape but subsidise their survival in poor years by providing artificial water and even supplementary food. Similarly, in moving to intensive agriculture, we aim to obtain much higher production per square metre than previously but do so by providing irrigation, fertilisers and a great deal of mechanical intervention through planting and harvesting. We replace natural but relatively inefficient and unreliable means of moving water and seeds across the landscape with artificial channels and mechanised planting. These developments are a necessary part of obtaining greater production per unit area of landscape, the principal societal goal for some regions, and this must be balanced with appropriate actions to meet other landscape goals such as the conservation of diversity. But these actions also reduce the intrinsic resilience of the agro ecosystem to deal with external shocks such as drought, a fact which we must therefore recognise and explicitly manage for.

Understanding the specific ways in which our ecosystems originally coped with the special features of the Australian biophysical environment is therefore essential in creating the appropriate management regime under which their natural resilience is replaced by management actions. Of course, the creation of such a management regime depends not only on knowledge of the biophysical system, but also on the social and political environment within which the knowledge is to be implemented, to which we now turn.

3. The social and political environment

Just as the biophysical environment of Australia has features which, whilst not being individually totally unique, in sum distinguish it from most other regions of the world where our lessons about drought management might be drawn, so too does its socio-political environment.

Looking back a matter of centuries, aboriginal inhabitants walked this continent relatively lightly, capitalising on years of plenty whilst coping with deprivation in years of drought (see Rose's chapter this volume). It was a classic low-input-low-output system, which had probably come generally into balance with variability in the biophysical environment over millennia of experimentation. People used large areas relatively conservatively. Various cultural rules were developed for protecting those resources that were critical in dry times through the times of plenty when people might have been tempted to overuse them. Whilst the demands of much higher population levels and higher expectations for standards of living mean that European-style, intensified land use is here to stay, there are lessons that can be learned from the ways in which that indigenous population created resilience in its interactions with the Australian landscape. To make such a case we must first look at the features of the Australian socio-political landscape which continue to distinguish it from other regions in the world.

First, and most obviously, the non-urban areas (whether just west of the dividing range or, more extremely, out in the arid zone) of Australia are relatively sparsely populated (see for example Haberkorn *et al* 1999). This reality has a whole series of downstream implications. Markets tend to be more remote or smaller, resulting in higher production and transport costs. Historically, remoteness has bred both a degree of self-reliance, but also a certain degree of disdain for the other end of the market chain, resulting in relatively poor feedback in terms of clients' concerns. The sparse population also means that it is relatively difficult for local groups to get together, to act together or to exchange information and understanding about how their systems work. In the past decade the Landcare movement has begun to evolve a missing level of local community governance in Australia, but with mixed success, particularly as regions become less densely populated. The limited communications technologies of the past have also tended to cause a disjunction between rural communities and the views of their urban peers, particularly in terms of growing concerns about environmental issues. Recent dramatic changes in communications and access to media are rapidly changing the situation but it has held true for the vast majority of Australia's European history.

Second, the Australian bush ethos has evolved with a curiously schizophrenic history. On the one hand, bush people have been fiercely independent and sceptical of the motives of central government. On the other hand, government has for so long promulgated and, with varying levels of effectiveness, implemented the intent of settling the empty continent, that people living out there have continually looked to the public purse to subsidise life in rural Australia. This, coupled with the almost mythological place that the outback has in the heart of urban Australia, enables a small rural electorate to have a disproportionate influence on the political process through the emotional ties of the urban populace. As we march into a period in which some regions are clearly moving in to a post-productivist future based on non-market values, whilst others stay with the productivist paradigm of the past (Holmes 1997), this schizophrenia is not declining.

Third, our federal political system creates additional oddities that any natural resource management institutions system must deal with. The responsibility for natural resources is vested in the individual states, which consequently replicate an immense amount of bureaucratic structure in order to implement laws for fundamentally similar purposes in each State. The Commonwealth, on the other hand, can only intervene through a limited number of instruments, including taxation and the provision of funding with cross-compliance requirements on some natural resource management issues. This is peculiar because, inasmuch as natural resource management issues need to be dealt with in a way which is sensitive to local conditions, this is needed at a regional scale. Ideally broad brush policy is consistent across the nation, however this needs to be negotiated at a federal scale. The state scale is intermediate and not particularly well-suited to meeting either purpose. The result is inconsistencies and conflicts, trivial or otherwise, across state boundaries, and enormous transaction costs for obtaining cross-border solutions. One would think that for many purposes a more satisfactory solution would be obtained through national policy-making with regional implementation.

To return to the example of the continent's earlier inhabitants, whilst we clearly cannot go back to the population densities and lifestyle hardships (by current standards) of those days, it is worth noting that their resilience to climatic variability in this land was sustained by an ability to obtain their needs from a large spatial area, by treating their core resources very conservatively, by the possession of intimate local knowledge, and through high social capital in the community. Each approach has its place in today's natural resource management. Policy can facilitate commercial or other arrangements that enable farmers to spatially hedge their exposure to drought, through cooperative arrangements with other regions; the significance of the many such arrangements that farmers already have for this are often overlooked. A balanced exploitation strategy in which core resources are not damaged remains crucial, and supporting social capital and local knowledge at the appropriate scale rather than imposing supposed national solutions remains highly relevant.

The remainder of this volume will be exploring issues of this type in greater detail. I complete this chapter by outlining some immediate general ways in which an understanding of the socio-ecological environments of Australia should inform future drought policy solutions.

4. Implications for drought policy

Drought policy must deal with at least two different types of issues. First, policy aims to facilitate the appropriate type of on-ground management; this implies that policy-makers must understand (in general terms, at least) what sort of management is required to respond to the biophysical conditions that managers are operating within. Second and equally important, policy instruments are themselves an institutional response to a problem, and policy-makers need to understand how to design the most effective types of institutions to obtain the intended on-ground outcomes.

Today, there is a growing body of theory and practice around the concept of learning communities and institutions (see for example Robbins *et al* 2002). Institutions (or more correctly the group of people operating within them) can only learn if they, first, are able to detect and attribute the positive and negative changes that their actions are causing and, second, have the knowledge, motivation and capacity to adjust their actions and structures in the appropriate way. This idea can be applied both at the scale of understanding on-ground management, and to the policy-making process itself. The goal is to have both land management practices and policy environments which are resilient, in the sense of being able to cope and evolve with changes and shocks from each of their respective external environments. In meeting such a goal, the specific local environmental conditions are crucial in understanding the local natural resource management responses; and the continental-scale diversity of environments together with their socio-political environments are crucial in informing the policy response. The remainder of this volume will address these issues in a variety of ways, but the following points are some examples of the general implications of the biophysical and socio-political environments that are particular to Australia.

4.1 LOCAL-REGIONAL SCALE ADAPTIVE SYSTEM ISSUES

At the local scale, there are a number of key environmental issues:

Australian agro-ecosystems generally have to cope with high levels of climatic variability at all scales from inter-annual to inter-decadal and longer; some regions, particularly across the centre of the continent, have an added element of uncertainty at the intra-annual level, that is not knowing when in the year rain is likely to come. Policy intervention needs to be aware of the many different local ways in which climatic variability plays out and affects managers across the continent.

Many environmental problems which have emerged from our management in Australian agriculture have very long lead times; they are hard to detect in their early stages and hard to attribute (for example Stafford Smith, Morton *et al* 2000). Inasmuch as policy seeks to create an institutional environment in which managers learn effectively how to cope with climatic variability, it must be recognised that this feedback loop is difficult at any time and misguided interventions can very easily reduce such tenuous signals as there are. Policy must also therefore facilitate the ability of managers to monitor the biophysical outcomes of their own actions, since these are hard to detect (and hence learn from) in a variable climate.

Many (but not all) Australian ecosystems are relatively unproductive compared to the expectations of the farmers using them; as a consequence, if the system is damaged there is limited capacity to invest in recovery. Understanding this aspect of system resilience is important, since it means that avoiding damage in the first place is more important in some systems than others (Stafford Smith, Morton *et al* 2000).

A core concept in resilience theory is that it is underlying 'slow' variables which are important issues in determining system resilience, not the superficial 'fast' variables. The 'fast variables' that humans depend upon in our day-to-day experiences are very real issues for short-term humanitarian aid, but confuse the strategic debate about sustainable natural resource management. Droughts bankrupt families when those families live on eroded landscapes with no stored capital, whether social or economic; the same 'drought' may hardly be noticed by a group of farmers with healthy pastures and low debt. In seeking to support long-term change towards sustainable livelihoods, it is essential to focus on the 'slow variables' which set this context (Stafford Smith and Reynolds 2002). In this sense, drought simply brings some underlying critical structural problems to a head.

Any policy solutions must address these local issues, but through a framework which can take account of some broader structural concerns.

4.2 STATE-NATIONAL SCALE ADAPTIVE SYSTEM ISSUES

At the policy formulation level there is another suite of more institutional issues.

The policy system needs to be able to learn and adapt to changes in the rest of the socio-ecological system just as much as individual farmers must. This may mean that it is important to design a system which can be sensitive to local, regional conditions. This is much easier in Landcare-type self-reliance type of institutional arrangements than through national taxation or reconstruction instruments. However, the national context for this system also needs to be designed in such a way that it can evolve constructively over time.

For the system to evolve over time, there must be feedback data on the actual outcomes (rather than inputs or even outputs) of policy actions, and a mechanism for responding to this feedback without loss of institutional coherence. It is therefore very important that the detailed instruments of any policy (which may need to change over time) are embedded within a well-articulated, broader policy philosophy which does not bounce around over time.

Climate variability is by definition a probabilistic beast. Policy success should not be judged on individual events but requires a long-term view, which does not sit easily with the electoral cycles if the policy interventions are too hands-on.

This chapter has outlined environmental factors which need to inform the philosophy and implementation of any drought policy in Australia. It has raised more questions than it has provided answers, but these may be sought in later chapters.

CHAPTER 2: CLIMATE AND DROUGHT IN THE SUBTROPICS: THE AUSTRALIAN EXAMPLE

JANETTE A LINDESAY*
*School of Resources, Environment and Society, The Australian National University,
Canberra, ACT 0200, Australia*

1. Introduction

The subtropical regions of the world lie between about 15-20 degrees and 40 degrees north and south of the equator, and include large parts of the US, South America, northern and southern Africa, the Middle East, most of Asia (including India and China), and the majority of the Australian continent. Together these areas are home to a large part of the world's population, many in developing regions, and encompass a considerable proportion of the global arable land area. They are also subject to some of the most marked year-to-year fluctuations of climate anywhere on earth.

Subtropical climates are distinguished by marked variations, particularly of rainfall, in both space and time. The high degree of both seasonality and interannual variability there is due to the location of the subtropics between the tropical zone straddling the equator and the middle latitudes further north and south. A variety of climatic mechanisms influence weather systems in these zones, and can interact in complex ways around the fringes of and across the subtropics. The central areas of the subtropics are also characterised by generally clear skies and relatively high summer temperatures. One consequence of a variable climate with distinct seasonality in rainfall and relatively high dry-season temperatures is an environment that is prone to periods of below-average rainfall and, in extreme conditions, to the prolonged rainfall deficits that produce drought.

As is true elsewhere in the subtropics, the environmental history of Australia has been discernibly shaped by climatic extremes of rainfall and temperature. The degree of adaptation of the biosphere to extremes of climate, and particularly to dry or drought conditions, is an indication that the general patterns of weather systems, climate and the seasons have existed in forms recognisably related to modern patterns since the continents reached their present geographic positions millions of years ago (Groves 1994). Human management with and adaptation to the climate and environment of Australia is a more recent development, with the arrival of aboriginal peoples dating from approximately 40,000-60,000 years ago and that of European settlers from the late eighteenth century (Allan and Lindesay 1998). The history of European responses to Australia's climate, in particular, and the environmental impacts of those responses have contributed to current landscape conditions and to many of the attitudes and values

*Material in this chapter was first published in Beyond Drought: People, Policy and Perspectives; Linda Courtenay Botterill and Melanie Fisher (eds). CSIRO PUBLISHING, Melbourne 2003. Reproduced by permission of the Publisher.

that we currently bring to managing with a variable climate. In the words of Blench and Marriage:

> Climate is often conceptualised as a series of shock events punctuating a background of acceptable variation. Shocks, such as floods, high winds and drought, are discontinuities that are sufficiently anomalous within the lifetime of observers as to be classified as unpredictable and life-threatening. The nature of the discontinuity is framed by the region's ability to cope. …Vulnerability to weather is a function of preparedness as well as of the event in itself. (Blench and Marriage 1999, p9)

The purpose in this chapter is to establish the climatological context for a discussion of drought in the subtropical regions of the world, using Australian examples of the nature and causes of climate variability and drought in the subtropics. The problem of defining drought is addressed, and drought definitions are considered across a range of sectors. The importance of using our understanding of past climatic variability, and drought in particular, to inform both current decision making and planning for the future is highlighted in a discussion of scenarios for possible future climate change in Australia, and what those changes could mean for the future of drought in this and other subtropical regions.

2. Defining drought

Drought is a normal and recurrent feature of climate; it has occurred everywhere on earth, but extended drought is more likely to be observed in regions of high interannual rainfall variability. The term 'drought' implies a lack of precipitation over an extended period of time, such that there is insufficient moisture for a particular region, land use or environmental sector. Moisture deficits are often exacerbated by high temperatures, low humidity and stronger than average winds, which increase evapotranspiration and the impacts of drought. One of the characteristics of drought, as distinct from most other natural hazards, is that drought conditions develop gradually and cumulatively as the balance between precipitation and evapotranspiration worsens progressively because of the failure of 'normal' rainfall. Thus the onset of drought conditions is often difficult to determine. The end of a drought may also be difficult to define, since the effects may linger for some time after precipitation increases again. Managing the impacts of drought is complicated by the fact that they are often widespread both geographically and across a range of economic sectors, and do not necessarily involve structural damage, making them difficult to quantify (for example, White *et al*, 1993; Glantz, 2000).

Drought is distinct from aridity, which is a permanent condition of low rainfall. However, the exact definition of drought is elusive; Wilhite and Glantz (1987) surveyed more than 150 drought definitions, and there are many more. Any useful definition of drought should be specific to the particular region and application under consideration. For example, regions where rainfall is generally frequent and relatively reliable may develop water deficit conditions after only weeks without rain, whereas in areas with pronounced dry seasons and high interannual rainfall variability, a season or more of below-average rainfall is required for a water deficit to develop.   In terms of applications, the water requirements of rain-fed agriculture and of hydrological systems are quite different, with the former more rapidly influenced by drier than normal conditions, particularly at critical times in the cropping cycle.   And areas of high water usage are more readily vulnerable to water shortage than those where water is used less intensively, highlighting the role that humans play in influencing and defining drought.

All definitions of drought are human constructs in that 'drought' or 'flood' are determined as departures from perceived 'normal' conditions; and the definition of what constitutes normality is itself not absolute, since it depends on the range of variability that has occurred during the period of record used to determine that normality.   For many people, human memory defines what is considered an extreme event. Scientifically, most climatic records cover no more than the last 100-150 years, so that definitions of extreme events have been derived in the context of century-long climate variability.   In recent decades, records have been extended backward in time using palaeoclimatic reconstruction techniques, which has allowed a broadening of the scientific view of what constitutes 'normal' climatic variability to millennia rather than centuries.   The perspective that is used in defining drought must depend on what is appropriate to the activity, time and place under consideration, so that exposure and vulnerability to drought impacts can be assessed.

2.1 DROUGHT DEFINITIONS

The *concept of drought* is best understood when definitions are given in general terms. Thus an agricultural definition of drought might be that:

> *Drought is a protracted period of deficient precipitation resulting in extensive damage to crops, resulting in loss of yield.*

The conceptual definitions of drought for applications such as hydro-electric power generation in New Zealand, or wildlife management in southern Africa, or urban water provision in the United Kingdom, would all be different.   One common factor in all drought definitions should be the incorporation of an understanding of climate variability. The conditions under which governments would provide financial assistance to those affected by drought impacts are those beyond what could be considered part of 'normal' climate variability and risk management.   Declarations of what might be termed 'exceptional' drought are then based on scientific assessments, bearing in mind that these are also not absolute.

Operational definitions of extreme events assist in identifying the beginning, end, and degree of severity of the event.   To determine the beginning of drought, operational definitions specify the degree of departure from the average of precipitation or some other climatic variable over some time period. This is usually done by comparing the current situation to the historical average, often based on a 30-year period of record. The threshold identified as the beginning of a drought (for example, 75% of average precipitation over a specified time period) is usually established somewhat arbitrarily, rather than on the basis of its precise relationship to specific impacts.

An operational definition for agriculture might compare daily precipitation values to evapotranspiration rates to determine the rate of soil moisture depletion, then express these relationships in terms of drought effects on plant behaviour (that is, growth and yield) at various stages of crop development. A definition such as this one could be used in an operational assessment of drought severity and impacts by tracking meteorological variables, soil moisture, and crop conditions during the growing season, continually re-evaluating the potential impact of these conditions on final yield. Operational definitions can also be used to analyse drought frequency, severity, and duration for a given historical period. Such definitions, however, require weather data on hourly, daily, monthly, or other time scales and, possibly, impact data (for example, crop yield), depending on the nature of the definition being applied. Developing a climatology of drought for a region provides a greater understanding of its characteristics and the probability of recurrence at various levels of severity. Information of this type is extremely beneficial in the development of response and mitigation strategies and preparedness plans.

2.2 PERSPECTIVES ON DROUGHT

Meteorological drought is defined on the basis of the degree of dryness (in comparison to some average amount of rainfall) and the duration of the dry period. Meteorological drought must be defined regionally because the climatic conditions that result in below-average precipitation vary; for example, drought could be defined in terms of days without rain in the wet tropics, but in terms of months or seasons without rain in the seasonally arid subtropics. Thus periods of meteorological drought are identified on the basis of relating actual precipitation departures from average amounts on monthly, seasonal, or annual time scales, as appropriate for the region under consideration.

Hydrological drought is defined in terms of the effects of below-average precipitation on water supply, *that is,* stream flow, reservoir and lake levels, and ground water levels and recharge rates). Hydrological drought and its impacts generally lag the occurrence of meteorological droughts because it takes some time for precipitation deficits to accumulate in components of the hydrological system, including stream flow, ground water and reservoir levels, and soil moisture. Land-use change can influence hydrological drought, or even cause reductions in infiltration and run-off in the absence of meteorological drought. Thus activities such as deforestation can have effects beyond the immediate area affected by meteorological drought.

Agricultural drought associates characteristics of meteorological and hydrological drought with agricultural impacts. Factors including precipitation shortages, differences between actual and potential evapotranspiration, soil moisture deficits, reduced ground water or reservoir levels are important in defining agricultural drought. Plant water demand depends on prevailing weather conditions, biological characteristics of the specific plant, its stage of growth, and the physical and biological properties of the soil. A good definition of agricultural drought should be able to account for the variable susceptibility of crops during different stages of crop development, from emergence to maturity.

The *sequence of impacts* associated with meteorological, agricultural, and hydrological drought further emphasises their differences. The agricultural sector is usually the first to be affected by drought because of its dependence on stored soil moisture which is depleted during extended dry periods. Continued precipitation deficiencies will affect other water users, with subsurface water resources often the last to be affected. A short-term (3-6 month) drought may impact only slightly on the latter, depending on the characteristics of the hydrologic system and water use requirements. When meteorological drought conditions diminish, soil water is replenished first, followed by stream flow, surface water reservoirs and ground water. Drought impacts may diminish rapidly in the agricultural sector because of its reliance on soil water, but linger for months or even years in other sectors dependent on stored surface or subsurface supplies.

Socioeconomic drought associates the supply and demand of some water-dependent economic good, such as water supply, animal feed, fish or hydroelectric power, with elements of meteorological, hydrological, and agricultural drought. It differs from these types of drought because its occurrence depends on identifying drought based on the temporal and spatial processes of supply and demand. Socioeconomic drought occurs when water supply is unable to meet economic demand because of weather-related factors. Since both demand and supply vary with time, it is important to identify possible convergence in the trends that could signal enhanced vulnerability to drought and so increase the incidence of socioeconomic drought.

2.3 EXPOSURE AND VULNERABILITY

The short-term meteorology (days to weeks) and longer-term climatology (months to years) influencing a region determine the *exposure* to drought—that is, the likelihood that atmospheric circulation anomalies leading to drought conditions will occur. Although these factors cannot be influenced by humans, some forewarning of their occurrence would allow informed planning. Considerable attention has therefore been given to developing seasonal climate forecasting (McKeon *et al* 1993; Nicholls 1997b; White *et al* 1999b). In contrast to exposure to drought, *vulnerability* is determined by human activity. The impacts of drought are dependent not only on the duration, intensity, and spatial extent of a drought, but also on the demands made by human activities and natural systems on available water supplies. Improved understanding of the past drought climatology of a region will provide critical information on the frequency and the intensity of historical events, and so may assist in planning for mitigating the impacts of future droughts.

3. The climate of Australia

The climate of the Australian continent is determined by the geography of the land mass, which extends from the tropics in the north, through the subtropics to the fringes of the midlatitudes in the south (Figure 1).   Each of these broad latitude zones is characterised by different dominant weather systems. In the tropics (from the equator to approximately 20 degrees South), easterly airflow and tropical disturbances (including tropical cyclones), and in some areas the wind reversals and distinct dry and wet seasons associated with the monsoon, dominate the rainfall regime. The subtropics (between about 20 and 40 degrees South) are dominated by anticyclones with slowly descending air masses and clear skies, disrupted by heat lows and tropical disturbances from the north and east in summer and by midlatitude cold fronts from the south and west in winter. In the midlatitudes (poleward of 40 degrees South), a succession of frontal low pressure systems moves across the fringes of the continent in the prevailing westerly airflow, each front bringing a sequence of weather conditions including warmer, dry air masses ahead of the front and cooler, moist air with the potential for rainfall and possible thunderstorms with the passage of the front (for example Colls and Whitaker 1990; Linacre and Geerts 1997; Sturman and Tapper 1996). These climatic zones play a fundamental role in shaping weather, climate and climatic variability over the region.

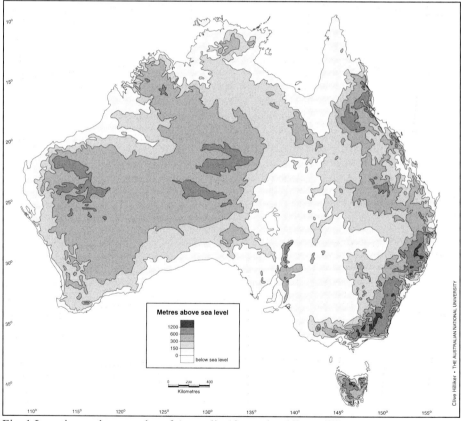

Fig. 1 Location and orography of Australia (drawn by Clive Hilliker, SRES, ANU)

Topography is also an important factor in determining regional climatic conditions in Australia. Although much of the continent has relatively low relief, the mountains of the Australian Alps in southwest New South Wales and northeast Victoria and the Great Dividing Range (or escarpment) inland of the east coast have at least a regional influence on climate. This is reflected in annual average rainfall that is higher on the windward (coastal) sides of the higher-altitude areas than it is on the inland sides; moist onshore airflow that is forced to rise above these mountains frequently produces cloud formation and precipitation, leaving moisture-depleted air to continue inland.

The large west-east extent of the continent means that inland Australia has generally drier conditions and more extreme temperatures than the coastal margins; the oceans generally have a moderating effect on temperature (due to their large capacity to absorb heat), and are often a source of moisture through evaporation. Areas in the interior of the continent, which lies largely in the subtropics, are remote from the sources of moisture-bearing winds, particularly tropical easterlies, crossing the coast. The interior also experiences greater extremes of temperature both diurnally (that is, the difference between daily maximum and minimum temperatures is relatively large) and seasonally (that is, between summer and winter) than do the coastal fringes. The diurnal temperature range is larger in the subtropics because the generally clear skies and dry air allow maximum heating of the surface and overlying atmosphere during the day, and maximum loss of heat by long wave radiation from the surface at night.

The result of Australia's geographical location and topography is a pattern of annual average rainfall and temperature with a general gradient from warm, monsoonal tropics to cool midlatitudes, and wetter conditions around the northern, eastern and southern coastal fringes (Figure 2). The interior is relatively hot and dry, with potential evapotranspiration greatly exceeding surface moisture availability; in central Australia near Alice Springs, for example, annual average rainfall is between 200 and 300 mm, while annual average potential evapotranspiration exceeds 1300 mm. These characteristics and their expression in the environment (in vegetation types, for example, which integrate the combined effects of moisture availability and temperature) have been summarised in classifications of climate, of which the best known is probably the Köppen Climate Classification (Figure 3). An important aspect of this and many other climate classifications is seasonality, since the timing of rainfall relative to the annual temperature cycle is important for the biosphere.

Seasonal variations in weather and climate occur principally as a result of the migration of the overhead sun between the tropics of Cancer and Capricorn, reaching a northern limit at the winter solstice in June and a southern limit at the summer solstice in December. The consequent winter expansion of cold Antarctic air masses and northward movement of midlatitude low pressure systems produces winter rainfall over southern Australia (for example, Perth, Figure 4c); and the expansion of warm, humid tropical air masses southward in summer brings tropical disturbances and moisture sources farther south over the continent (for example, Figure 4b) (Colls and Whitaker 1990; Hobbs 1998; Sturman and Tapper 1996). Across northern Australia the seasonal migration of the zone of maximum convergence of airflow and cloud formation between the northern and southern hemisphere tropics leads to the wind reversal (moist south-easterly in summer, dry north-westerly in winter) associated with the monsoon (for example, Darwin, Figure 4a). These atmospheric circulation characteristics mean that there is some degree of seasonality in rainfall almost everywhere in Australia. Differences between January (summer) and July (winter) rainfall are clearly greatest in the monsoonal tropics and in the southwest winter rainfall region, and seasonal temperature differences are large in the subtropical interior, the southwest and the southeast. The influence of topography on both temperature and rainfall is particularly evident in south-eastern Australia.

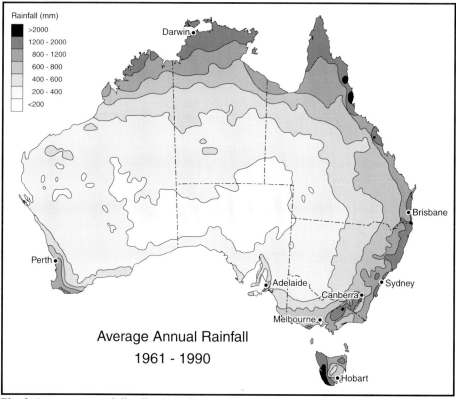

Fig. 2 Average annual distribution of rainfall totals in Australia, 1961-1990 (after Australian Bureau of Meteorology, www.bom.gov.au)

The seasonal characteristics of Australian climate are average conditions, however. As is typical of the global subtropics, much of the continent is subject to a marked degree of interannual climatic variability (Figure 5). The weather and climates of both the tropics and the midlatitudes may be affected by large-scale circulation fluctuations from year to year that can affect the location, frequency and intensity of particular types of weather systems. Any changes in these two zones will affect the subtropics, which are thus particularly prone to interannual variability as expressed in extremes of rainfall and temperature. The impacts of year-to-year fluctuations in rainfall, in particular, can be large in regions characterised by distinct dry and wet seasons; a failure of rain during the normally wet season means that no effective falls can be expected until the start of the next wet season. This can result in prolonged periods of below-average rainfall, and drought.

4. Changing rainfall seasonality

Recent research on changing seasonality in Australia rainfall (Lindesay and Johnson 2003) has highlighted the spatially varying nature of long-term fluctuations in monthly rainfall across the continent. Probably the most important aspect of these fluctuations is the changes that have occurred in rainfall in some months of the year, while rainfall in other months has changed little during the more than 100 years of meteorological record. At Sydney in south-eastern Australia, for example, since 1860 there has been an increase in rainfall in January and a decrease in July, both of the order of 30% of median monthly rainfall for those months. The result is that at Sydney, and at many other places in Australia, rainfall seasonality is not fixed, but varies between periods of well-defined, high-amplitude seasons and periods when rainfall seasons are poorly defined and the transitions between seasons may be blurred. The fact that the seasons are not fixed in either amplitude or timing, and that the start and end of the wetter season in subtropical Australia has varied by at least a month during the last 100 years or so, adds to the complexity of identifying drought. These fluctuations in rainfall are not random, however, and are apparently related to some of the most significant large-scale climatic causes of low-frequency rainfall variability in Australia.

5. Causes of climate variability

Perhaps the most widely recognised large-scale influence on interannual rainfall variability in the global subtropics is the El Niño Southern Oscillation (ENSO) (see, for example, Allan *et al* 1996; Glantz *et al* 1991; Hobbs *et al* 1998), which is particularly important in modulating rainfall variations from year to year across much of eastern Australia. The ENSO phenomenon is the largest known interannual fluctuation in the ocean-atmosphere system. It centres on the tropical Pacific Ocean, although its characteristic patterns of atmospheric pressure, winds and temperature extend into the Indian Ocean region, and its impacts are near-global. ENSO involves a suite of interlinked anomalies in both atmosphere and ocean, most obviously seen in sea surface temperature variations across the tropical Pacific Ocean.

The lowest air pressures tend to be co-located with the highest surface temperatures in the tropics, so that on average there is a low pressure centre (the Indonesian Low) over the 'maritime continent' area north of Australia where sea surface temperatures are high (Figure 6). The South Pacific Anticyclone (a high pressure centre) is associated with the cold sea surface temperatures along the South American coast in the eastern South Pacific. The southeast trade winds blow across the tropical Pacific between these two pressure centres, from the northern fringe of the anticyclone towards the low upwards in the atmosphere. This energy is then transferred eastward (across the Pacific), westward (over the Indian Ocean) and southward at altitudes of about 10 km, before the air sinks towards the surface in the semi-permanent anticyclones of the subtropics. These large-scale overturning features of the general circulation of the atmosphere are known respectively as the Walker (east-west or zonal) and Hadley (north-south or meridional) circulations. Any disruption to the 'normal' state of the atmosphere and ocean across the tropical Pacific Ocean can thus lead to changes in winds and rainfall-producing weather systems throughout the tropics and into the subtropics and midlatitudes, via changes to the large-scale atmospheric circulation.

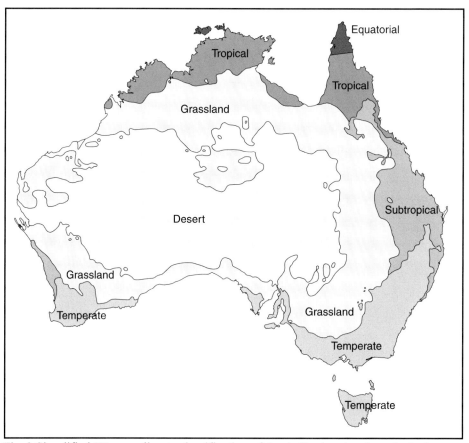

Fig. 3 Simplified Köppen climate classification of Australia showing the principal bioclimatic regions (after Australian Bureau of Meteorology, www.bom.gov.au)

Changes in the gradient in sea-level pressure between the two principal Pacific basin pressure centres are known as the Southern Oscillation, and are used as a measure of the status of ENSO using a variety of indices. The best-known of these is the Tahiti-Darwin Southern Oscillation Index (SOI) (Allan *et al* 1996). The SOI is calculated as the difference in monthly average sea-level pressure between Tahiti (in the vicinity of the South Pacific Anticyclone) and Darwin (in the Indonesian low pressure area). Each month's pressure value is standardised to remove the seasonal cycle, which would otherwise dominate the record. The SOI calculation used in Australia follows the method of Troup, which provides typical SOI values between +20 (La Niña events) and

-20 (El Niño events); the pressure gradient measured by the SOI is a continuum, however, and is more often closer to 0 (normal) than in either extreme state. Time-series of the Tahiti-Darwin SOI can be calculated from 1876 (Allan *et al* 1996) and have been correlated with rainfall in many parts of the world, including Australia. Other, more accurate, indices of ENSO activity have been developed using shorter data series; amongst these are the Multivariate ENSO Index (MEI) and the Niño 3.4 Sea Surface Temperature Index (based on sea surface temperatures in an area of the central equatorial Pacific), both of which can be used to track ENSO since the 1950s.

A

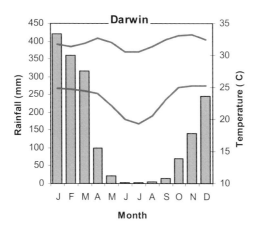

B

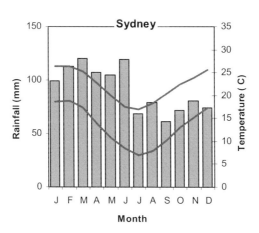

C

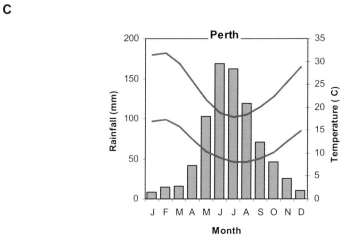

Fig. 4 The average seasonal cycles of rainfall (left) and temperature (right) at (a)
Darwin (tropical), (b) Sydney (subtropical east coast), and (c) Perth (subtropical west
coast) (data from Australian Bureau of Meteorology, www.bom.gov.au)

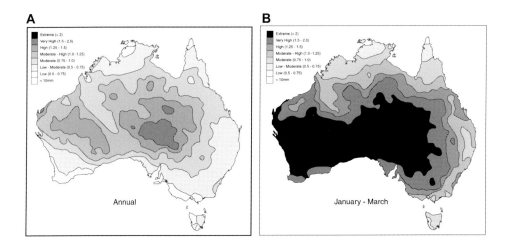

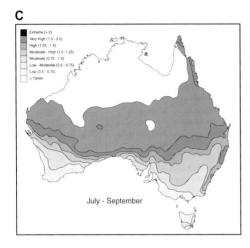

Fig. 5 Average interannual rainfall variability over Australia, 1961-1990: (a) annual, (b) summer (January-March), and (c) winter (July-September). Numbers in the key are values of variability, calculated as $90p - 10p/50p$, where $90p$, $50p$ and $10p$ are the 90th, 50th and 10th percentiles respectively (after Australian Bureau of Meteorology, www.bom.gov.au)

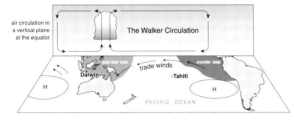

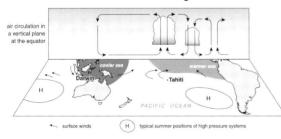

Fig. 6 Oceanic and atmospheric features across the equatorial and southern Pacific Ocean region during normal (top) and El Niño (bottom) conditions (after Climate Diagnostics Center, NOAA, www.cdc.noaa.gov)

During the El Niño extreme of ENSO, above-average sea surface temperatures extend from the South American coast into the central equatorial Pacific, and temperatures over a large area may rise to more than 6°C above normal. At the same time, sea surface temperatures around the northeast Australian coast are often cooler than normal. This represents a significant change to the normal east-west temperature gradient across the tropical Pacific (which is generally warmest in the west, i.e., north of Australia, and coldest in the east, i.e., off the South American coast). Sea-level pressures rise across the western equatorial Pacific and northern Australia and fall in the area of the south Pacific Anticyclone, leading to a decrease in the gradient of pressure and thus a decline in strength of the tropical south-easterly trade wind flow (Figure 6). The principal atmospheric low pressure area and convergence zone in the tropics, normally located over the warm waters north of Australia in summer, moves eastward to the vicinity of the international dateline where the area of above-average sea surface temperatures has developed. Tropical convection and moist easterly onshore airflow diminish over Australia, and parts of eastern Australia are thus generally sunnier, drier and warmer than normal during El Niño events (Figure 7a). These events recur on average every 3-4 years (Allan *et al* 1996), although the return period varies: for example, there were four successive El Niño years in the period 1991-1995, followed by another event in 1997-98. The most recent El Niño event occurred in 2002-03, and was associated with one of the most severe drought periods on record in parts of eastern Australia.

Wetter than average conditions typically occur in eastern Australia during La Niña events (Figure 7b), when tropical Pacific sea surface temperatures are colder than usual. There is a tendency in the climatological record for La Niña conditions to follow El Niño events, although this is not always the case. This contrast could account, at least in part, for Australia's reputation as a land of droughts and flooding rains.

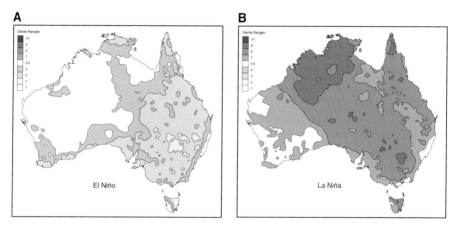

Fig. 7 Average annual rainfall deciles for (a) 12 El Niño years, and (b) 12 La Niña years El Niño years are: 1905, 1914, 1940, 1941, 1965, 1972, 1977, 1982, 1987, 1991, 1994, 1997. La Niña years are: 1910, 1916, 1917, 1938, 1950, 1955, 1956, 1971, 1973, 1974, 1975, 1988 (after Australian Bureau of Meteorology, www.bom.gov.au)

Despite its well-described behaviour and general impacts, ENSO is highly variable; no two El Niño events have the same evolution, intensity or impacts. This makes precise prediction of ENSO impacts particularly difficult, as is exemplified by the moderate (although spatially coherent and physically meaningful) overall correlations between ENSO and rainfall in Australia. It also means that indices of ENSO activity such as the SOI must be used cautiously in a predictive context. ENSO is not the only influence on rainfall in Australia, and not all droughts can be ascribed to El Niño events. Other contributing factors include the Pacific Decadal Oscillation (Power, Tseitkin *et al* 1999), the Indian Ocean Dipole or Dipole Mode Index (Saji *et al* 1999), and the Antarctic Circumpolar Wave (White 2000c).

Recent research has highlighted the potentially important role of the low-frequency fluctuations of the Pacific Decadal Oscillation in modulating ENSO influences on Australian rainfall (Power, Casey *et al* 1999). The phenomenon, which has characteristic patterns of sea surface temperature across the northern Pacific Ocean, varies slowly on decadal to multi-decadal timescales (Figure 8). When the Pacific Decadal Oscillation is positive, ENSO events are more likely to impact moderately on Australian rainfall; when the Pacific Decadal Oscillation is negative there is an increased likelihood of ENSO events impacting strongly in Australia. The Pacific Decadal Oscillation, which was positive between around 1980 and 2000, has since entered a negative phase. Thus the background climatology for the 2002-03 drought season, with an El Niño event occurring in a negative phase of the Pacific Decadal Oscillation, may have assisted in establishing the conditions for severe impacts of a relatively moderate El Niño event (as measured by the SOI and MEI) in Australia.

Understanding the causes of climatic variability, such as the significant role of the oceans and sea surface temperature variability (as exemplified by ENSO and the Pacific Decadal Oscillation), enhances our ability to unravel the complex interactions between the impacts of natural fluctuations in the system and those impacts that may be ascribed to human activity. Physical understanding of the nature of climate fluctuations and their impacts is also contributing to seasonal climate forecasting, which can be a useful tool for land- and water-resource managers and in agriculture. Knowledge of forthcoming drought conditions will not permit those conditions to be avoided, but does allow appropriate management decision making to minimise drought impacts.

6. Aridity and drought in Australia

As has been shown, much of Australia lies in the semi-arid subtropics, with average annual rainfall totals below 350 mm per year. Long-term average rainfall is calculated as the arithmetic mean of annual rainfall at each location over a period of time, usually at least 30 years (the most recent World Meteorological Organisation climatological averaging period is 1961-1990). However, in regions of high interannual variability such as the subtropics, the long-term average may not reflect the actual nature of rainfall in any particular year (Figure 9). Identifying drought can be particularly difficult in these areas.

7. A brief history of Australian drought

Rainfall deficits and drought are a recurrent feature of the climate of Australia. On average a severe drought occurs somewhere in Australia approximately once in 18 years; actual return intervals vary between 4 and almost 40 years. The characteristics of individual droughts are just as variable, with some rainfall deficits accruing over years while other droughts are short and intense. It is also common for some regions to record good rainfall while others are in drought. Droughts in Australia affect agriculture through crop and stock losses, and are often associated with severe bushfire seasons, dust storms and soil loss, and environmental degradation in general.

The most widespread and/or intense droughts, such as those of 1982-83 and 2002-03, have an immediately discernable impact on the Australian economy, and have environmental and economic consequences beyond the end of the meteorological drought

The droughts that have made a particular impression in Australia include:

- 1895-1902
- 1914-15
- 1937-1945
- 1965-68
- 1982-83
- 1991-1995 and
- 2002-03.

Some of these drought periods were related to the occurrence of El Niño events (for example 1982-83), and in some cases large parts of the country were affected (for example 1901-02). But in other cases drought was more localised, or occurred in the absence of an El Niño event under the influence of other large-scale climatic factors. The following brief descriptions of significant drought periods in Australia during the last 100 years or so are based largely on material from the Australian Bureau of Meteorology *Climate of the 20th Century* project (Bureau of Meteorology 2003).

The Federation drought of *1895-1902* was amongst the most environmentally damaging types of drought, with one or two drought years following a prolonged period of generally below-average rainfall. Across much of Australia the rainfall pattern was marked by dry spells through the years before Federation in January 1901, particularly in 1897 and 1899. Following some rain in many areas in 1900-01, dry conditions became established across eastern Australia, with Queensland, New South Wales and Victoria in drought by mid-1902 (Figure 10a). The drought finally broke in December 1902. The impact on agriculture was severe, particularly on the wheat crop and on livestock; much of Queensland had been drought-affected for eight years by the end of 1902.

The next major drought was that of *1914-15*, which was memorable not only for its intensity but also for the fact that much of Australia was affected, beginning with South Australia, Tasmania, Victoria and New South Wales. Unusually, both the eastern states and the south of Western Australia simultaneously suffered dry conditions that led to the total failure of the national wheat crop in 1914. There was a strong El Niño event in 1914, and severe bushfire conditions occurred in south-eastern Australia during the dry summer that year. Although this 18-month drought was not uniformly dry either in time or in space, it was the worst drought of the century in some areas. It ended in the winter/spring of 1915.

A

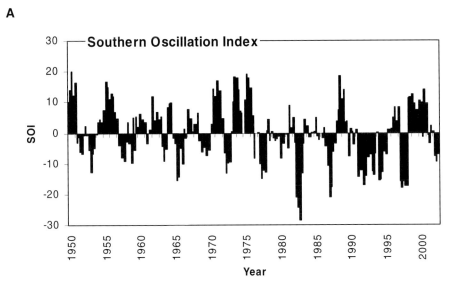

B

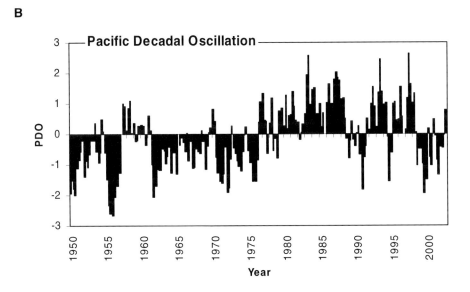

Fig. 8 Three-month seasonal values of (a) the Tahiti-Darwin Southern Oscillation Index (SOI), 1950-2002; and (b) the Pacific Decadal Oscillation (PDO), 1948-2002 (data from Climate Diagnostics Center, NOAA, www.cdc.noaa.gov)

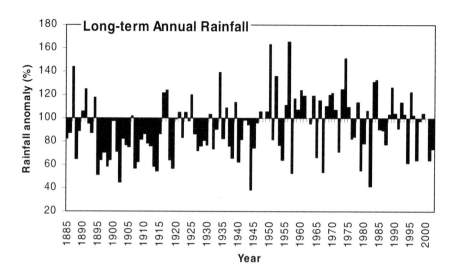

Fig. 9 Annual rainfall anomalies at Fairlight Station (35.2°S, 148.9°E), near Canberra, 1885-2002. Anomalies are expressed as percentage departures from the 1961-1990 long-term mean (*i.e.* 120% = 20% above-average rainfall, and 80% = 20% below-average rainfall) (data by the Australian Bureau of Meteorology, www.bom.gov.au)

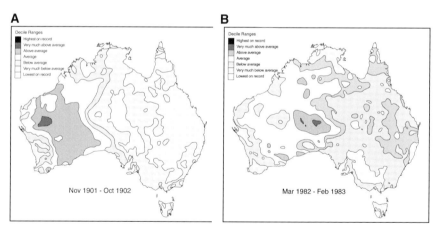

Fig. 10 Average annual rainfall deciles for two El Niño drought years: (a) November 1901-October 1902, and (b) April 1982-February 1983 (after Australian Bureau of Meteorology, www.bom.gov.au)

The late 1930s and early 1940s were another period of generally below-average rainfall, with significant droughts occurring in *1937-38, 1940-41 and 1943-45*. Severe drought conditions began in New South Wales, Queensland, Victoria and some parts of Western Australia in 1937, spreading to South Australia in 1938. Conditions in the southwest had serious impacts on the wheat crop, and in January 1939 Victoria experienced the 'Black Friday' bushfires. Following good rains during 1939, drought conditions returned in 1940 with a strong El Niño event; this was one of the driest years on record across southern Australia. Good rainfall was recorded in the second half of 1941 and in 1942, but the drought resumed in 1943 and, in many parts of Victoria, South Australia, New South Wales and Queensland, continued to mid-1945 and beyond.

The period from *1957 to 1968* was comparable to the Federation and 1937-45 droughts in severity and areal extent. A complex succession of drier and wetter years marked this period, and the most intense rainfall deficits occurred in various parts of the country at different times, including in northern and central Australia. The eastern states were affected most during 1965-1968, with severe drought in south-eastern Australia in 1966-67 accompanied by bushfires in Tasmania and dust storms in South Australia. The drought ended late in 1968.

The *1982-83* drought was one of the most intense and widespread on record in Australia (Figure 10b), and has been described as having the worst overall impacts of droughts in the twentieth century (Allan and Heathcote 1987; Allan *et al* 1996; Glantz *et al* 1987). This was a year in which an El Niño event developed rapidly and particularly strongly across the Pacific basin during mid-1982, and the SOI reached its lowest values in a century. Drought conditions occurred across most of Australia, but particularly in the eastern half of the country (with the exception of northeast New South Wales and southeast Queensland). A number of dust storms were generated as strong winds stripped dry, unvegetated topsoil from extensive areas in south-eastern Australia, and in February 1983 dangerous fire-weather conditions led to several large bushfires burning across South Australia and Victoria in the devastating Ash Wednesday fires of 16 February. Agricultural production was cut by an estimated 10% as a result of the 1982-83 drought, with an estimated cost to Australia from all drought-related losses of around $3 billion.

The next significant droughts occurred during *1991-1995*, a period characterised by the unusual (but not unprecedented) occurrence of four consecutive El Niño years. Conditions were particularly dry in Queensland, northern New South Wales, and parts of central Australia during this period. The northern wet season failed in the Northern Territory in 1991-92, and Queensland remained dry, although some rain occurred further south across south-eastern Australia. By January 1994 extended drought conditions across New South Wales contributed to severe bushfires around Sydney. Although some rain fell in autumn 1994, the development of stronger El Niño conditions led to increasing drought across the country from mid-1994 into 1995; the drought that year is estimated to have reduced agricultural production by 8%. This protracted drought period ended with good rains in late 1995 and 1996.

Most recently, *2002-03* saw one of the most widespread and severe droughts on record across much of Australia, this time accompanying a relatively moderate El Niño event. In this case the Pacific Decadal Oscillation was in a negative phase (as it had been in the 1960s, for example), contributing to the magnitude of the ENSO impact in Australia. Severe rainfall deficiencies in almost every state were exacerbated by well above average temperatures, particularly in parts of Queensland and New South Wales. Maximum temperatures averaged 1.59°C above normal in 2002, and were 1.65°C higher than average in March-November 2002 (a record for Australia) (Karoly *et al* 2003). Drought conditions from April 2002 to January 2003 resulted in a severe bushfire season in areas of Victoria, Tasmania, New South Wales and the Australian Capital Territory, with the Canberra fires of 18 January 2003 the second most costly in Australia's history.

Drought has been a recurrent feature of the Australian environment throughout the period of instrumental record—that is, since the mid to late nineteenth century. Its occurrence has been variable in both space and time, but an increased understanding of the basic climate state and the nature of climate variability, and of climatological factors influencing drought occurrence and severity (for example ENSO and the Pacific Decadal Oscillation), has allowed the development of seasonal climate outlooks and advisories that are useful for natural resource and agricultural planning and management (Nicholls 1997b; White *et al* 1999b). An additional factor that must be taken into account, however, is the possibility of climate change.

8. Climate change and the future

Against the background of a climate system characterised by a high degree of variability on a range of timescales, it is now acknowledged that some degree of human-induced change in climate is occurring and will continue to occur into the future (Houghton *et al* 2001). The changes that have been observed over the last 100 years or so, particularly in air temperature and, in more complex ways, in rainfall, are due to a combination of natural variability and increasing concentrations of carbon dioxide and other heat-absorbing gases in the atmosphere. The amounts of these gases present in the atmosphere have risen exponentially since the 1850s, largely as a result of increasing combustion of fossil fuels, agricultural practices, and land-use change (including deforestation). The result of increasing atmospheric concentrations of the greenhouse gases is a general increase in near-surface air temperature. Over the same period, industrial processes have led to an increase in the amount of sulfate particles in the atmosphere, which have a cooling effect on near-surface air temperatures and so moderate greenhouse warming to some extent (Houghton 1994; Houghton *et al* 2001).

The overall response in the climate system has been a globally averaged warming of approximately 0.6°C since 1900. The observed changes in global air temperature and carbon dioxide in this period exceed the maxima reached during at least the past 500,000 years (IPCC 2001). In Australia the average temperature across the continent has risen by 0.7°C between 1910 and 1999, with much of the warming occurring in the second half of the twentieth century. The warmest year on record across the continent was 1998. As has occurred in many other regions, Australian night time minimum temperatures have increased more rapidly than daytime maximum temperatures, leading to a reduction in the difference between daytime and night time temperatures (the diurnal temperature range). Over the same period, rainfall, which is inherently more variable in space and time than temperature, has not shown any clear trend in the continental average. There are, however, some identifiable trends in regional rainfall (Bureau of Meteorology 2003).

A significant international scientific focus on possible future greenhouse gas emissions scenarios (Houghton *et al* 2001) has resulted in projections of global average warming of 1.4-5.8°C by 2100 (relative to 1990), with an average warming rate of 0.1-0.5°C per decade. Simulations of the climate system using complex Global Climate Models and these global warming projections allow the development of scenarios for average temperature and rainfall change across Australia in the future (CSIRO 2001). Regionally specific projections are being developed using finer resolution (limited area) models such as the CSIRO Atmospheric Research DARLAM model (for example, Whetton *et al* 2001). It is not possible to predict conditions in individual years, since natural climatic variability will continue to affect the interannual fluctuations of climate in unpredictable ways.

Overall, annual average temperatures are expected to rise by 0.4-2.0°C across much of Australia by 2030, with slightly smaller increases in the south and larger increases in the northwest. By 2070, temperatures are expected to be 1.0-6.0°C higher than in 1990. The largest increases occur in summer, and a greater frequency of very hot summer days (with temperatures exceeding 35°C) and a concomitant decrease in the frequency of winter days with sub-zero temperatures are predicted across the continent.

The projected changes in rainfall across Australia are more spatially variable, and carry higher uncertainties than do the temperature projections (CSIRO 2001). Annual rainfall changes between -20% and +20% are predicted by 2030, with the biggest reductions in the southwest and parts of the southeast, but little change in the tropics across the northern parts of the continent. By 2070, rainfall may have changed by as much as 60% in some areas. Seasonal patterns of rainfall change are complex: in winter and spring the trend is towards drier conditions, whereas in summer and autumn, some areas become wetter, while others are drier. A further important aspect of possible future rainfall conditions is the predicted increased frequency of extreme rainfall events (CSIRO 2001).

Future scenarios for Australian climate in the twenty-first century are thus characterised by higher temperatures and, in some areas, by drier conditions, particularly in the southern and western parts of the continent. There is, however, marked seasonal variation in the rainfall projections. Research on ENSO behaviour under greenhouse warming scenarios indicates that there is unlikely to be any significant change in ENSO; that is, the interannual fluctuations of sea surface temperature and atmospheric circulation will continue to produce conditions that will result in El Niño and La Niña events in future. However, these events will be impacting on an environment potentially more vulnerable to extreme conditions due to the changes in climate described here. It seems probable that these changes, with adjustments in both the long-term mean and in overall variability, will result in a general shift towards increased frequency of high-temperature, extreme-rainfall events. Human-induced climate change has been suggested as the cause of the combination of severe rainfall deficits and high temperatures (with associated soil moisture loss and vegetation drying) that characterised the 2002-03 drought across large parts of Australia (Karoly *et al* 2003). The possibility of such drought impacts recurring in future raises implications for land and water resource management and for agriculture that will have to be addressed in policy, planning and management.

9. Summary and conclusions

This chapter has established the climatological context for a discussion of drought in the subtropics, and particularly in Australia. Subtropical climates are characterised by clearly defined rainfall and temperature zones and seasons, influenced by the location of these regions between the tropics and the midlatitudes. The most significant features of climate in the subtropics, and in Australia in particular, are the strong seasonality of rainfall, the degree of change in seasonality during the last 100 years or so, and the relatively large interannual rainfall variability. Year-to-year fluctuations in climate are due to often complex combinations of large-scale influences including the El Niño Southern Oscillation, the Pacific Decadal Oscillation, the Indian Ocean Dipole, and others. The result of the nature of the Australian climate, and of that elsewhere in the subtropics, is that conditions of rainfall deficit or drought are relatively common.

Defining drought is not a simple matter, however, and a range of definitions has been considered. Different components of human and natural systems respond to a rainfall deficit in different ways, often complicating the assessment of the beginning and end of drought conditions. There is no doubt, however, about the impacts on Australia of the most severe droughts of the last century; these are evident in an overview of the drought history of the continent. The importance of using our understanding of such past climatic variability, and drought in particular, to inform both current decision making and planning for the future is highlighted in a discussion of scenarios for possible future climate change in Australia. If rising global temperatures and changing climate may result in an increased frequency of high-temperature and extreme-rainfall seasons, the implications for land and natural resource management should inform policy formulation and management planning.

CHAPTER 3: INDIGENOUS WATER PHILOSOPHY IN AN UNCERTAIN LAND

DEBORAH BIRD ROSE
Centre for Resource and Environmental Studies, The Australian National University, Building 43, W K Hancock Bldg, Canberra, ACT 0200, Australia

1. Introduction

Right across Australia, indigenous people hold water to be sacred. Their management of their use of freshwater, including care and restraint, constitutes their essential adaptation to this driest of inhabited continents. Geographically, my analysis is concentrated mainly on the arid and semi-arid zones, as together they comprise approximately 80% of Australia's land mass; here the first principles of water philosophy and practice can be examined in high relief. I weave the sacred geography of water with people's pragmatic knowledge and use of water. To try to separate these aspects of water knowledge would be to misrepresent the power and beauty of indigenous water praxis.

Indigenous Australians learned to understand water in order to adapt themselves to it— to its unpredictability, its capacity to support life, its dangers, and its hidden places. Not only did they acquiesce and adapt to the water conditions of the Australian continent, they also enhanced the capacity of water to sustain life. Acceptance of the water conditions of any given territory was not passive non-action, but rather was and is an active way of working with water's own action. Practices of care involve relationships between people, water, and all the living things that depend on water, and thus entail ethics. I use the term philosophy to refer to this domain of knowledge, ethics, and practice.[1]

2. 'TEK'

In the past few decades scholars have been paying ever greater attention to 'traditional ecological knowledge'. Although many of us are dissatisfied with the term, it does signal three important aspects of the knowledge developed and sustained by indigenous people (Hunn 1993, pp13-15). First, it is 'traditional' in the sense of having been developed, tested, assessed, and transmitted over many generations. It is knowledge pertinent to the local area; it has a history, and it has enabled people to survive. In Australia more than 3,000 generations of human occupation (over a period greater than 40,000 years) underpin contemporary indigenous knowledge.

---

[1] The use of the culture-specific term 'philosophy' has the danger of distorting the understanding of Indigenous life practice in its inclusive and interactive dimensions. At the same time, however, it most clearly corresponds to the conjunction of root paradigms, ethics, and practice that I am aiming to describe and analyse in this chapter.

L.C. Botterill and D.A. Wilhite (eds.), From Disaster Response to Risk Management, 37–50.
© 2005 *Springer. Printed in the Netherlands.*

Second, TEK concerns ecology. Scholars began by eliciting taxonomies, developing the methods and data that enable us to grasp the enormous diversity, as well as the many important similarities, in human environmental perception and knowledge. Increasingly, however, TEK is focussed on process. In Lewis's (1993, p9) terms, TEK looks to

> the understandings that people have of environmental systems and the networks of cause and effect therein. A part of these understandings involves a people's perceptions of their own roles within environmental systems; how they affect, and how they are affected by, natural processes.

Third, TEK involves knowledge. The western love affair with technology led to classifications of the world's peoples on the basis of tool-types, and people with relatively small tool kits were classed as 'primitive'. The logical next step was to assume that their knowledge also was minimal. Australian Aboriginal people were hunter-gatherers; they have been subjected to this form of misapprehension, and thus it is always important to emphasise that for mobile peoples, tool kits were limited by what people could carry with them. Knowledge, however, is carried in the mind. Whatever the limitations of the human mind, there is no one-to-one relationship between technology and knowledge. Research continues to confirm that Australian Aboriginal people's knowledge of the natural world is both sophisticated and extremely detailed.

Within this developing field of indigenous knowledge, water has until recently been something of a poor cousin. No doubt influenced by the western tendency to take water for granted, we have overlooked some of the most important questions that could be asked. And yet, as Tom Greaves (1998, p36) contends, 'global scarcities of freshwater have a particular relevance to indigenous societies, to their future and to their universal cultural rights'. His argument is that for people whose identity and future is linked to place, cultural survival depends on their homeland being habitable. If their waterways are diverted, damaged, or destroyed, their right to cultural survival is thereby threatened. He sets his analysis within the context of competing social rights to water. Along with social issues of water use and allocation, there is the struggle to conserve freshwater ecosystems against a range of biological and chemical invasions that threaten to destroy them (Daiyi *et al* 2002).

3. Law

For the Aboriginal people of this challenging continent, water does not happen by chance, but rather exists through the creative action of Dreaming beings. The term 'Dreaming' connotes both creation and connection. It refers to the beings who made the world, and it further refers to the continuing process of life's coming forth in the world, thus referencing both original and on-going creation. Dreaming creation is termed Law in many Aboriginal languages and cultures. The late Mussolini Harvey, a senior Law man whose homeland was in the islands of the Gulf of Carpentaria off the coast of the Northern Territory of Australia, explained to non-Aboriginal people the meaning of 'Dreaming'. It is one of the most eloquent explanations in existence. He begins:

White people ask us all the time, what is Dreaming? This is a hard question because Dreaming is a really big thing for Aboriginal people. In our language, Yanyuwa, we call the Dreaming *Yijan*. The Dreamings made our Law or *narnu-Yuwa*. This Law is the way we live, our rules. This Law is our ceremonies, our songs, our stories; all of these things came from the Dreaming. (in Bradley 1988, p xi).

Rock holes, soaks, wells, rivers, clay pans, water-holding trees, billabongs, springs and the like form part of the subsistence geography of country and invariably part of the sacred geography as well. The tracks and sites of Dreaming significance link surface, subsurface and aerial sources of fresh water. Indigenous knowledge of underground water is not as well studied as the knowledge of surface water. The indications are that there is a great deal for outsiders to learn. The Karajarri people of the Kimberley, for example, distinguish between the water which flows under the ground toward the coast, and the freshwater streams that feed the springs. The first type is a 'deep salty, pressurised stream that bubbles to the surface'; the second is freshwater. It arises on land, and also under the salt water in the sea (Lingiari Foundation 2002, p13).

Rainbow Serpents and other large snakes are regularly associated with permanent water and with connections between the subsurface, surface, and aerial waters (see for example Radcliffe-Brown 1930). Everywhere they are dangerous, and everywhere they are life-giving. Across much of Aboriginal Australia, where there is permanent water, there is the Rainbow Serpent—in the rivers, springs, and aquifers. In western New South Wales, in the country of the Ngiyampaa people, Steve Meredith spoke about *Wawi*, the Rainbow Serpent:

> This country was made by the ancestors. *Wawi* the Rainbow Serpent came up through the springs, he came from Nakabo springs, Ngilyitri country. Wherever he travelled he left ochre to show where he had been. The springs were entry and exit points. He came out of the earth, travelled along its surface, and then went back to the earth. *Wawi* travels, and is still there. We know he's still there. (quoted in Rose *et al* 2003, p62)

The Rainbow Serpent is also claimed as the creator of many surface rivers and lakes. In many parts of Australia it is said to have created water systems, including the lakes and rivers. According to the late George Dutton, of western New South Wales:

> That's why the river Darling is like a snake's track, where they travelled along and bored a channel to make a river. They bored out lakes: they coiled around and scooped out the sand into sand hills. And they made channels to drain into the lakes such as creeks. To make the water run this way and they rose up. Now this was done right through the Australian land. (quoted in Beckett 1958, p106)

The Dreaming beings are ancestors of groups of living things that include both human and non-human members. The stories tell of how they travelled in both their human form and in the form of the species which they would become, and how they changed over to become fixed in contemporary species and in their human counterparts. Some of the stories concern water's living things—the Barramundi Dreaming track, Crocodile track, or Turtle track, among others. Most of the species of animals, and many species of plants, have specific origin in Dreaming actions. Thus, the Crocodile Dreaming, for example, is ancestral to contemporary crocodiles and to the crocodile people who are its human relations. Similarly with turtles, and similarly with barramundi, and so on. To be born into an Aboriginal family is to be born into relationships with the land and water of one's parents' country and to be born into relationships with a variety of terrestrial and aquatic living things.

Throughout most of Australia the most plentiful and most reliable water sources are also likely to be sites in which plants and animals are protected. Water is life for everybody, not just for people. Dreamings created relationships that structure obligations of care, and that constitute webs of reciprocities within the created world. These relationships are localised. Dreamings established countries. A country is small enough to accommodate face-to-face groups of people, and large enough to sustain their lives; it is politically autonomous in respect of other, structurally equivalent countries, and at the same time is interdependent with other countries. Bonds of mutual life-giving are focussed in country, and countries are connected through Dreaming tracks to form regions.

Every country has its own permanent and ephemeral waters. No country is without water as that would leave people dependent on others, but in some areas, of course, water is very scarce. Water in Aboriginal Australia exists within a system of rights and responsibilities that is usually referred to as ownership. Groups of people belong to and 'own' their country, including their water. Rights and responsibilities are vigorously defended. In the words of Marcia Langton, one of Australia's leading indigenous scholars:

> individual rights and responsibilities arise from the wider mytho-
> geographical bodies of knowledge, and…these rights and responsibilities
> of individuals in relation to…waterscapes…are jural in nature. (Langton
> 2002, p45)

Indigenous people shared resources, as well as defending their ownership rights and responsibilities. The general rule, articulated in simple and eloquent terms was and is: 'always ask' (Myers 1982). The rule identifies the right of the owners of country to say yes or no; 'always ask' articulates the right and responsibility of owners to make managerial decisions about the use of their own country.

3.1 FLEXIBILITY

Aboriginal people spaced themselves across the continent in densities that reflect the rainfall of a given area. Much like the Anglo-European settlers who now cluster in the well-watered coastal areas, Aborigines, too, sustained higher densities where this was possible, and sustained scattered low-density populations where that was all that the water could support. Each group defined by language ('tribe' in much of the literature) belonged to its own territory (cluster of countries), and each tribe comprised approximately 450 people. Thus in the arid regions of the deserts, tribal territories were very large, and along the coastal areas, tribal territories were correspondingly much smaller. The people held group size relatively constant, and adjusted the size of the territories to suit (Tindale 1974, p31 and p111).

Tribes were grouped together into larger clusters forming cultural blocs. Peterson (1976) studied the cultural and natural areas of Australia and found that the tribal groups and major drainage systems showed a high degree of correlation. At a finer level of resolution, in areas where there is a network of creeks and rivers, ownership of country tends to flow with the water. Groups who share adjacent junctions on a larger river, or whose countries come together at a watershed between two creeks, share close relations in marriage and trade, and often share a language. In the inland, linguistic boundaries tend to align with ecological zones.

Across the arid zone, water is relatively scarce and is also unpredictable. Murrell (1984, p327) describes desert rainfall as 'notoriously unreliable' with 30-50% deviation from the average. Tindale (1974, p31) found in his research in some of the most arid regions that the size of territories could be as large as 300-400 kilometres in diameter. Each territory had at least one permanent water source that was retained strictly for the group, and other permanent water sources that were shared with other groups. The sharing of water sources was not necessarily harmonious, but it was a necessity, and it seems to have been understood as such in areas where rainfall is both seasonally unpredictable, being influenced by ENSO, and also locally unpredictable. In the western desert much of the rainfall comes in the form of thunderstorms—one area will be drenched, while an immediately adjacent area will remain completely dry. This happens sporadically, and without pattern, with the exception of the rare cyclonic depressions that may bring rain to a large area. Unpredictability is a relative concept, and requires a corresponding notion of predictability. Dick Kimber (1997, p12) had the good fortune, as he expresses it, to travel in the deserts and to learn from indigenous people who were thoroughly at home there. He notes that 'whitefellas' labelled desert conditions as 'very changeable', whereas to the people who belonged there, conditions were in fact 'entirely natural'.

Drought, according to Tindale (1974, p68-9), 'is the great and extraordinary hazard of desert life; one that comes with such frequency as to engrave its pattern on the lives of most generations ...' He himself encountered people struggling to get themselves to food and water in the wake of drought. In these cases, the disruptions caused by settlers' use of the land, and the spread of feral animals such as rabbits, clearly exacerbated the effects of drought. Nevertheless, it is reasonable to suppose that prior to European disruptions there must have been occasions when some people in some areas were unable to survive drought.

Indigenous people's main adaptation to uncertainty was to develop social ties that enabled people to move to resources as they became available. The social organisation of sharing was utterly essential. Gould's work on risk and sociality offers key insights. He maintains that desert people lived by 'chasing water'. One of their major strategies, therefore, for 'minimizing the risks in an inherently risky environment is to establish and maintain multiple, long-distance kin sharing networks that enable people to move freely to better favoured areas during drought.' In discussing the constant sharing behaviour and the networks of obligation, he concludes that 'sharing relationships among these people are too important to be left to sentiment' (Gould 1982, p73). A strong effect of this system is that while asking is obligatory, there are intense pressures toward sharing, not least of which is the knowledge that there will be times when others will have to reciprocate.

One aspect of the genius of this system of countries is to give people inalienable rights without inhibiting their flexibility. Social relations cross-cut the boundaries of rights and responsibilities without obliterating or undermining them. On the one hand, countries were under the control of the people who belonged there and who, through creation, bore responsibility for the country and its living things, including water. People protected their own country from strangers out of necessity as well as love of country. Their ongoing subsistence depended on control of resources, and this is a matter of life and death. For example, one cannot plan to rely on a certain place that is a source of fresh water, only to arrive and find that somebody else has used it all up. Use and access rights must exist and be enforceable. Control of knowledge was, and is today, a key form of defence. 'Intellectual property' rights to knowledge constitute the heart of territorial integrity and thus of sustainable inhabitation.

Water knowledge of Walmadjari people of the Great Sandy Desert is well documented. Walmadjari country is at the margins of northern monsoonal influence. The rainfall is uncertain in amount, but is relatively predictable by season. Tindale provides a list of ten terms for water sources in Walmadjari language that include soaks, clay pans, rock holes (two types), permanent springs (two types), river pools, creek pools, brackish or salty water, and the sea (Tindale 1974, p63). There are also numerous terms for types of rain (Mangkaja Arts Resource Agency 2003, p20; see also Simpson 1997). There is, further, a major distinction between permanent and ephemeral waters. The permanent waters are called 'living water' (Lowe 1990, p11), and they constituted the core area of a country.

Pat Lowe, whose life and work with the late Jimmy Pike documents Walmadjari knowledge, offers more precise insight. Her elegant book *Jilji* is illustrated with her own photos and with Jimmy Pike's art work. She writes that the most important water sites are jila—waterholes where the water is permanent. Often these waterholes are small and insignificant to look at, but they are of extreme significance to people because they support life when other ephemeral water sites have failed. Lowe writes:

> A jila is living water, a place where water can be found at any time of the year, and it was close to a jila that people used to live late in the dry season when all other sources of water had dried up. ... Next in importance was a jumu. Like a jila, a jumu has to be dug out, and after a wet season it would be difficult to tell the difference between the two types of waterhole, which can look much the same. But desert dwellers had to know which waterhole was a jila and which was a jumu because a jumu would eventually dry up. Unlike a jila, it could not be relied upon

far into the dry season… The desert people had to pay careful attention to the pattern of each year's rainfall in order to be able to predict whether a particular jumu would yield water. To make a mistake would mean, at best, a long, parched journey to the next reliable water source: at worst it would mean death from thirst. (Lowe 1990, p 11, 14)

The middle and upper reaches of the Victoria River in the Northern Territory, where I have conducted large amounts of research (Rose 2000; Rose 2004 in press), is savannah country ranging from semi-arid to arid. Like the Great Sandy Desert, the Victoria River region is affected by the annual monsoons, but without the predictability of the coastal regions. In this savannah zone there is an annual period when the heat and humidity build to extreme levels, and the country becomes astonishingly desiccated in advance of the rains which may or may not arrive. This is a period of stress for humans, other animals, and plants. People have identified a large, detailed, and localised set of indicators that tell about the escalating heat and implicitly tell about the status of ephemeral waters. These indicators depend on temporal concurrence of spatially separated events. Signs tend to cluster around trees which bear seeds. When the pods of the bauhinia tree (*Lysiphyllum cunninghamii*) turn the deepest red, for example, the message is that the hottest weather has arrived. The implication is that ephemeral waters are going or gone, depending on their longevity. Further south, beyond the zone of bauhinia trees, a grevillea (*Grevillea dimidiata)* makes a similar announcement. Underground water is also marked by signs. Certain plants, for example, signal that there is water underground. Here the concurrence is spatial as well as temporal. Other signs of spatio-temporal concurrence, such as the behaviour of birds, indicate the presence of water.

Another significant aspect of knowledge is recognition of patterns based on sequence. One of the people I have been learning from in NSW explained that when the wild and damaging bush fires swept through the country in early 2003 the old Aboriginal people were happy. They expressed an idea that to me seems an excellent example of ENSO wisdom. They told the younger people that it always goes like this: after the fire the rain will come. Fire announces the return of rain after a drought, and cleans up the country so that a fresh cycle can begin.

As these brief examples indicate, knowledge tends to be highly localised and extremely detailed. The point made by Walker et al (Walker *et al* 1995, p86) in respect to dryland rivers is equally valid for rain in the arid zone: both require a long time 'to demonstrate their repertoire of temporal behaviour.' For humans, age is thus a critical factor in knowledge. People build up knowledge of events out of experience, and that experience could make the difference between life and death. Jake Gillen (pers comm) refers to elderly and mentally active people as 'ecological encyclopaedias'.

Jones and Meehan discuss the significance of age in Arnhem Land, where water is generally thought to be relatively predictable. There was:

a broader scale of unpredictability. A wet season might be late, or, even worse, might fail altogether; a cyclone might tear great strips through the forests … There was no means of long-term forecast, but there was what one might call long-term 'backcast', that is, a deep knowledge of what had happened in the past, and what the solutions were then…. Which wells had remained fresh, when all the others had gone salty or had dried up, or conversely, which dune ridges had remained above flood water when all of the surrounding countryside was inundated. Such knowledge usually resided with old people, and it was at these points of crisis that the accumulated knowledge of the entire group of people became a decisive factor in their survival. (Jones and Meehan 1997, p18)

In these storied landscapes, where Dreaming tracks connect water sites and other remarkable geographical features, the songs and stories carry life and death knowledge that is owned and vigorously protected by the people who have the responsibilities for the country, the water, the stories, the Dreamings, and the section of song that maps the country. An excellent example of the role of song comes from the Simpson Desert. This is one of the world's great sand ridge deserts. Much of it receives annual rainfall of less than 13 cm in an 'average' year, and summer temperatures may exceed 50°C (Shepard 1999, p x). Dick Kimber (1986) recorded irreplaceable information on the Simpson Desert in his life history of Walter Smith. As a young man, Walter Smith travelled the Simpson with one of the last of the truly great desert bushmen, Sandhill Bob. In the book *Man from Arltunga*, Walter Smith recounts some of the knowledge that Sandhill Bob drew on to keep them oriented, and to move them from water to water in country where the sandhills not only shift, but all look very much alike.

Walter described how people were taught to look at a particular tree and take their bearings off the branches of the tree, to count the sandhills and to remember how many have been counted, to navigate by the stars, to know where people had built cairns to mark underground water, to know which water could be stored in a water bag and which could not. Here as elsewhere, the Dreamings created water, and the songs tell about its presence. Sandhill Bob sang the songs of the country as they travelled through it along the Dreaming tracks (Kimber 1986, pp76-85).

4. Rainmaking

Many powerful Rain Dreaming tracks start up in the desert. The cultural logic suggests that the best knowledge of rain is to be found amongst people who are most in need of it. These Dreaming tracks are powerful and dangerous. They connect Rain sites with Lightning sites and other storm and flood sites. Among the powers associated with these sites is the power to call up rain. Rain-making rituals have been documented across the whole of Australia, along with rituals for diverting rain when it is unwanted (for example, when people are in the midst of ceremony).

Westerners have tended to be extremely sceptical of the efficacy of rain-making rituals. Such matters fall well outside the bounds of scientific explanation, and the evidence suggests that their effectiveness, like the rain itself, is unpredictable. On the other hand, the literature is replete with instances of rain-making rituals that are effective. An early and interesting account is Mrs Langloh Parker's comments on a woman in her employment who was skilled in rain-making. This woman erected two posts in Parker's garden to protect it from drought. She writes that this woman:

> was going away for a trip. Before going she said, as she would not be able to know when I wanted rain for my garden, she would put two posts in it which had in them the spirits of Kurreahs [Rainbow Serpent] …. (Parker 1905, p47).

Mrs Parker adds that her garden was well-watered.

Much more recently, Dick Kimber's research with some of the leaders of Rain Dreamings enabled him to observe the work of these men as they redirected rain so that ceremony could continue. He writes that their power was 'perceived as such that they can not only enhance the prospects of rain, but also direct its course or, if needs be, cause it to stop' (Kimber 1997, p10).

Perhaps the most interesting thing about rain-making is the scale. Whatever people may or may not have accomplished, it seems clear that they were not aiming to transform landscapes; they did not seek to make the deserts bloom all the time, or to keep the rivers always in flood. The philosophy underlying rain-making parallels that of other forms of water adaptation and enhancement: people sought to enhance water's capacity to nourish life without seeking radically to alter the water conditions of their country or, cumulatively, of the continent.

4.1 SOCIALITY

Lowe describes a pattern of water use that was widespread across the continent, saying that 'jumu were important because there were so many of them scattered through the desert, and they enabled people to travel far from their main jila for hunting and food gathering during much of the year' (Lowe 1990, p14). This pattern is that after rains, people used the ephemeral water sources, leaving the main waterholes to recover. As the ephemeral waters dried up, people moved back to longer lasting water, eventually settling around the main waterhole, unless disaster (such as lack of food) forced them to seek out neighbours.

The use of dryland rivers of the interior shows another side of this pattern—the enhanced sociality afforded by significant widespread rains. Keith Walker refers to the ENSO-affected Australian dryland rivers as a boom-and-bust ecology (Walker et al 1997, p64). The same could be said of the human life associated with these rivers. Out in the interior of the Simpson Desert there are sacred sites and other meeting places where people gathered episodically as the rains permitted. Around the north and west quadrants of the Simpson desert, for example, Arrernte people's countries are organised along the major rivers which flow into the desert and disappear there. These rivers are dry most of the time, but with rain they start to flow, and at the same time, the desert begins to flower. Arrernte people followed their rivers out into the desert, singing songs of the country and of the Dreamings who travelled these same routes; they sang of flowers, colours, and butterflies, of love and desire, and of all the sudden joyful life. They followed the tracks of Dreamings who brought ceremony, and groups that might only see each other during the best rains (which might be years and years apart) met up at major Dreaming centres for regional ceremony, trade, marriages, initiations, dispute resolution, and to enjoy the temporary abundance of the flourishing desert. As the desert dried up, they retreated back to the permanent waters that sustained them most of the time.[2]

On the east and southeast side of the desert the water regime is different, and people's adaptation was correspondingly different. In the area that includes the Diamantina River and the country west and northwest into the heart of the desert, people developed an ingeniously flexible system. The Diamantina has its headwaters in a northern, better watered region of Queensland where the monsoon influence gives a seasonality to flood pulses (although they vary in size and duration). It takes months for the floodwaters to travel from their origins in the northeast to Lake Eyre in the southwest, and they periodically extend the width of the Diamantina up to 500 kilometres (White 2000b, p67). As these channel country rivers travel, they sustain a richly abundant wetland ecosystem (Kingsford *et al* 1998).

In the southeast sector of the Simpson, as in many other parts of Australia, indigenous people dug and maintained wells. This region was characterised by permanent human habitation. Called *mikiri* in Wangkanguru language, wells were dug into existing soaks in the centre of the low-lying swales between the dunes. The well was made by digging or tunnelling into a soak that lay above a clay layer and so kept the water relatively close to the surface. Some of the wells went as deep as seven metres. Each well was a named and storied site. People had duties of care toward their home wells that included keeping the well free of siltation and rubbish (Hercus and Clark 1986).

People of the inside desert maintained close social links with their neighbours along the Diamantina River; although they owned different countries (riverine and sand dunes), they visited back and forth in response to the varied abundances afforded by rain in the desert and water in the river.[3] They may also have taken short-term refuge in each other's areas, depending on the vagaries of rain and flood. Here the genius of flexibility finds extraordinary fruition, as people gained the opportunity to respond to the water flows of two separate rainfall systems. The desert water flow is episodic (thunderstorms, with rare influence from more distant systems). The riverine water flow has its origins in the monsoon and ENSO-influenced rains of Queensland that make their way down the Diamantina.

---

[2] Information gained from Arrernte people's evidence in land claims which were presented under the *Aboriginal Land Rights (NT) Act 1976*.

[3] Information gained from Wangkanguru people's evidence in a land claim presented under the *Aboriginal Land Rights (NT) Act 1976*. Evidence from nearby river systems suggests that this type of adaptation is at least 13,000 years old, and perhaps much older (Smith et al 1991)

4.2 ENHANCEMENT

There are many, many references to the care that people took to ensure that their water sources remained as beneficial as possible. The *mikiri* wells, for example, were kept clean and clear; the sides were propped up with wooden structures. In many accounts, the care and cleaning of waterholes is linked to ritual, and it is clear that some of the ritual is secret and sacred, and thus is not available for contemporary discussion.

Pat Lowe vividly describes the work of digging out a waterhole:

> At the waterhole, the men are still digging. Already they have reached damp sand. There is only enough room in the deepening hole for one to work, so they take turns on the shovel. As the pit sinks lower and the man digging has to throw sand upwards, he abandons the shovel and digs with a billy can… Now the sand in the bottom has turned to grey mud.

> The man in the pit squats down and, after consulting those who know this place the best, scoops a smaller, deeper hole to the east of his feet: the Ngapa Mil, or eye of the jila. Water starts to seep in. The sides of the inner well are crumbling, and the man calls for [hunks of desert grass] to line the mil [eye].… The well man curves the grasses into an arc and, stooping, presses them into the seepage hole. Everyone else stands on the heaped sand above, watching the water slowly rise. A cup is called for, and a woman produces an enamel mug. The man below tosses out the first scoops of water, dark and murky. When the cleaner water has seeped up to replace it, he swirls away the surface scum and immerses the mug. This second water, still only inches deep, is grey and gritty, but he drinks deeply, refills the mug and passes it up. The water goes from hand to hand. It is cool and fresh and tasting of the earth. (Lowe 1990, pp136-37).

Other forms of life enhancing care include protection for plants and animals. Peter Latz, a botanist who has carried out extensive work in Central Australia, notes that the most sacred/protected places are likely to be places where a number of Dreamings meet up or cross over. He describes them this way:

> …there's a lot of dreaming trails which cross over, these are really important places. They are so sacred you can't kill animals or even pick plants. And of course you don't burn them. You might burn around them in order to look after them. (Latz 1995, p70)

Most of the 'really important places' focus on water. The restraint enjoined as respect for sacred sites enhances the capacity of such sites to serve as refugia in times of drought.

Some of the most vivid evidence of indigenous curation and protection of water comes through contrast with settler profligacy. Peter Latz has had the opportunity to study the process in Central Australia, and he states bluntly:

> The Arrernte people ... have important sacred sites where lots of
> Dreamings meet up with each other. These places were like ... the biggest,
> the most wonderful cathedral in Australia. And, of course, they were also
> the best places for recolonisation. There's a place called Running Waters,
> the best waterhole in central Australia, which was an absolute sanctuary.
> The waterhole runs for about four miles. Pelicans breed in it. It is now
> utterly stuffed! It was the very first place that white people came in and
> unwittingly put all their cattle. In other words, it's as if the whites came
> up here, found the cathedral and then went and shat on the altar! (Latz
> 1995, p84)

Similarly, Veronica Strang states that on the west coast of Cape York, mining is having
large impacts on water (Strang 2001, p211). Aboriginal elders evaluated the pollution of
water caused by mining companies as 'a poisoning of the rainbow—a flow of alien
substance into the lifeblood of the community' (Strang 2001, p223). One of the elders
explained that:

> pollution from upriver is seen as potentially disastrous not only to
> resources, but "everything":
> "...It will all be gone, finish. All the fish, all the animals,
> everything finish." (Strang 2001, p222)

5. Conclusions: philosophy in practice

Water is part of the sacred geography of people's homelands; it is part of creation,
connection, and an ethic of responsible care. Water knowledge is integral to the broader
domain of ecological knowledge, and water is invariably linked to life. 'Living water'
conveys the sense of water having its own life, and also of offering life to others.
Human understandings of water elaborate both aspects: water in its own presence or
absence, and water in its connectivities across land forms, between earth and sky, and
among living things.

I can summarise the discussion thus far by formulating several principles through which
Indigenous people's adaptation to extreme uncertainty has enabled sustainable
inhabitation across the millennia. They include:

- People manage people, not water
- Knowledge of water's 'behaviour' underpins action; water's capacity is enhanced
  wherever possible
- Responsibilities are focussed in country, confined to groups and defended by groups
- Flexibility is sustained across countries through social relations
- The necessity of sharing is balanced by the dictum 'always ask'
- People make extensive use of their range and opportunities by using ephemeral
  waters first and saving permanent waters as the waters of last resort
- Knowledge is coded and transmitted in song and story and through personal
  memory; experiential knowledge is highly valued

- Pattern recognition is focussed on several types of pattern: temporal concurrences, spatio-temporal concurrences, and sequences within cycles

The significance of water as a life-giving substance is metaphysically elaborated in many regions in the sacred geography of clan wells. The relationship between person and clan well is described eloquently by Mr Bulun Bulun of the Ganalbingu people of the Arafura wetlands in Arnhem Land. He speaks of the Ral'kal (sacred waterhole):

> Ral'kal translates to mean the principal totemic or clan well for my lineage. Ral'kal is the well spring, life force and spiritual and totemic repository for my lineage… It is the place from where my lineage of the Ganalbingu people are created and emerge. It is the equivalent of my 'warro' or soul. Djulibinyamurr is the place where not only my human ancestors were created but according to our custom and law emerged, it also the place from which our creator ancestor emerged…. Djulibinyamurr is my Ral'kal, it is the hole or well from which I derive my life and power. It is the place from which my people and my creator emerged. (quoted in Langton 2002, p49).

The relationship between water, humanity, creation, Law, and life is recursive and dynamic. It signals an intersubjective ethic within which self and other exist in relationships of connectivity such that each depends on the other. Every living thing, including water, is enmeshed in numerous such relationships, and the effect of this deep interconnection is to undermine the idea that self and other can exist in an either-or relationship. Connectivity, conceived so profoundly and pervasively, avoids the problem of managing narrow self-interest. Where the well-being of self and other is mutually interdependent, then the ethic of care of both self and other is embedded in both philosophy and practice (see Rose 1999). Such an ethic, focussed on water, brings all of life into fluid and recursive interdependencies of mutual concern. The emphasis is on life: on life's capacity to flourish, on life's capacity to keep on flourishing from one generation to the next. Taken to this deep level, water philosophy leads people into forms of respect and care that are characterised as holiness.

According to Yolngu elders of coastal Arnhem Land in the Northern Territory, 'In the Yolngu world view, water is the giver of sacred knowledge, all ceremonies and lands. Whether it's fresh or salt, travelling on or under the land, or in the sea, water is the source of all that is holy' (Ginytjirrang Mala 1994, p5).

In recent years, water has also become a powerful metaphor for the encounter between indigenous knowledge and Anglo-Australian scientific knowledge systems. Yolngu people speak of the estuarine mingling of fresh and salt waters, saying that they can productively meet and mix, but that in the larger perspective each also retains its own integrity (Yunupingu 1991).

As the model of water is applied to cross-cultural knowledge, the metaphor suggests that both indigenous knowledge and western knowledge are integral knowledge systems. Each has its own integrity, and yet they can interact productively. The implications of indigenous water philosophy for drought management are most provocatively expressed in the proposition that it is people who need to be managed. This proposition goes against the western technological orientation and managerial mind set, but if we are to accomplish a long-term shift toward a 'culture of permanence' (Wilson 2002, p23) we must start taking seriously such provocative ideas.

CHAPTER 4: LATE TWENTIETH CENTURY APPROACHES TO LIVING WITH UNCERTAINTY: THE NATIONAL DROUGHT POLICY

LINDA COURTENAY BOTTERILL
National Europe Centre, 1 Liversidge Street (#67C), Australian National University, ACT 0200, Australia

1. Introduction

From the time that they arrived in Australia, Europeans regarded drought as an aberration, a break with the 'normal' pattern of climate, and its onset was considered to be a natural disaster. Until 1989, governments responded accordingly through Commonwealth-State natural disaster relief arrangements which treated drought in a similar manner to other disasters such as cyclones, earthquakes or floods. With the removal of drought from these disaster arrangements in 1989, this view of drought as disaster was questioned in policy circles and a view emerged that drought was a normal part of the farmer's operating environment and should be managed like any other business risk.

In 1992 the Commonwealth and State governments agreed on a national drought policy based on principles of self-reliance and risk management and a package of programs was put in place to support farmers as they improved their risk management skills. The policy also introduced the concept of 'exceptional circumstances' to cover events of such severity that they were beyond the scope of good risk management.

This chapter describes the development and evolution of Australia's National Drought Policy and tracks the changes that have occurred over its first decade of operation and the policy challenges that continue to face policy makers working within its framework.

2. Drought as a disaster

To the Europeans who arrived in Australian in the late eighteenth century, Australia's uncertain climate presented a major challenge. An early report back to the British House of Commons into the prospects for agriculture in the colony of New South Wales noted the 'uncertain climate' and suggested that the future of the colony

> will be that of pasture rather than tillage, and the purchase of land will be made with a view to the maintenance of large flocks of fine-woolled sheep; the richer lands, which will generally be found on the banks of the rivers, being devoted to the production of corn, maize and vegetables. (Bigge 1966, p92)

L.C. Botterill and D.A. Wilhite (eds.), From Disaster Response to Risk Management, 51–64.
© 2005 *Springer. Printed in the Netherlands.*

In spite of these early warnings, a strong rural sector developed and set the foundations for Australia's prosperity. The adage that 'Australia rode on the sheep's back' was an accurate picture of the Australian economy through the second half of the nineteenth and first half of the twentieth century. From the time of the arrival of the Europeans in 1788 until 1989 when a national drought policy was introduced, Australia experienced severe droughts in the 1840s, 1895-1902, 1914-15, 1937-1945, 1965-68 and 1982-83 (see Lindesay, this volume). Under the Australian constitution, natural disaster relief is a responsibility of State governments and until 1939 the Commonwealth government had little involvement in disaster relief. Drought was one of the factors considered in debate leading to the *Wheat Growers Relief Bill (No 2) 1934;* however, it was not the sole focus of the legislation. The first explicit disaster relief intervention by the Commonwealth government was in 1939 when a grant of £1000 was made to the Tasmanian government to assist that State with recovery from bushfires. The Commonwealth then became increasingly involved in providing disaster, including drought, relief on an *ad hoc* basis. This included special purpose legislation such as the two *States Grants (Drought Assistance) Acts* of 1966. In 1971 the division of responsibility between the Commonwealth and States for disaster relief was clarified with the establishment of a formula setting out the relative contributions of the two levels of government (Snedden 1971). The arrangements left responsibility for the declaration of natural disasters with the State governments and the States were required to fund disaster relief up to a predetermined threshold, except for the relief of personal hardship or distress. Once this threshold was reached the Commonwealth Government would contribute funding on the basis of the formula. These arrangements have been amended a number of times since their inception but the basic structure remains.

By 1989, expenditure on drought relief was dominating the natural disaster relief arrangements and in April that year the Commonwealth Minister for Finance announced that drought was to be removed from the arrangements. This decision followed the tabling in the Commonwealth Parliament of a leaked Queensland government report that had uncovered considerable rorting of the scheme (Walsh 1989, p302). There were also allegations that the Queensland Minister had used his discretion to overrule a departmental decision to deny drought relief to a relative of the State Premier (Koch 1989). In addition to these problems with the administration of the scheme, it was also becoming increasingly untenable, in light of improved understanding of Australia's climate patterns, to argue that drought was a disaster and not a normal feature of Australia's climate.

Following the Government's decision to remove drought from the natural disaster relief arrangements, the Drought Policy Review Task Force (DPRTF) was set up to:

1. identify policy options which encourage primary producers and other segments of rural Australia to adopt self-reliant approaches to the management of drought;
2. consider the integration of drought policy with other relevant policy issues; and
3. advise on priorities for Commonwealth Government action in minimising the effects of drought in the rural sector. (DPRTF 1990, Vol.1, p2)

The DPRTF argued that 'drought is a relative concept, not some absolute condition. It reflects the fact that current agricultural production is out of equilibrium with prevailing seasonal conditions.' It went on to state that 'managing for drought is about managing for the risks involved in carrying out an agricultural business, given the variability of climate. Drought represents the continuing risk that seasonal conditions will not be adequate to sustain agricultural activity' (DPRTF 1990, Vol 1, p3). The Task Force rejected the construction of drought as a disaster and recommended that a national drought policy be implemented 'as a matter of urgency' (DPRTF 1990, Vol 1, p21). The review team identified its focus clearly—'The concept of risk management is central to the philosophy of this review' (DPRTF 1990, Vol 1, p14)—and it set out its vision of the role of both government and farmers in achieving a sound drought response. It argued that any government assistance should:

- be provided in an adjustment context
- be based on a loans-only policy
- permit the income support needs of rural households to be addressed in more extreme situations. (DPRTF 1990, Vol 1, p18)

Throughout 1991 and 1992, Commonwealth and State Ministers for agriculture debated an appropriate response to the Task Force report. The mechanism for these discussions was the Australian Agriculture Council (later the Agriculture Council of Australia and New Zealand), a Commonwealth-State ministerial consultative arrangement which had been in place since 1934, and its Standing Committee of Officials.

In 1992 the Senate set up an inquiry into an appropriate government response to the Task Force report. The inquiry expressed concern at the length of time that had elapsed since the Drought Policy Review Task Force had reported and called on Ministers to agree a national drought policy (Senate Standing Committee on Rural and Regional Affairs 1992, p xi). The Senate inquiry endorsed a self-reliant approach to drought and, like the DPRTF, rejected the reinstatement of drought in the natural disaster relief arrangements.

3. The 1992 National Drought Policy

Against this background of a general consensus on the direction of drought policy in Australia, Ministers announced in July 1992 that they had reached agreement on the National Drought Policy, based on principles of self-reliance, risk management and an acceptance that drought was a natural feature of the Australian climate. It was agreed that 'in circumstances of severe and exceptional drought' an appropriate response would be considered that would 'not compromise the principles and objectives' of the National Drought Policy (AAC 1992, p25). Its objectives were to

- encourage primary producers and other sections of rural Australia to adopt self-reliant approaches to managing for climate variability;
- facilitate the maintenance and protection of Australia's agricultural and environmental resource base during periods of increasing climate stress; and

- facilitate the early recovery of agricultural and rural industries, consistent with long-term sustainable levels. (ACANZ 1992, p13)

The policy set out clearly the responsibility of the government and farmers in implementing this new approach. Farmers were asked 'to assume greater responsibility for managing the risks arising from climatic variability' while the government would 'create the overall environment which is conducive to this whole farm planning and risk management approach'. However, governments recognised that special assistance would be needed in the event of severe drought and they undertook in these circumstances to 'act to preserve the social and physical resource base of rural Australia, and provide adjustment assistance in the recovery phase' (ACANZ 1992, p13).

3.1 MAIN COMPONENTS OF THE POLICY

The main mechanism for delivering support to drought-affected farm businesses was through the Commonwealth government's Rural Adjustment Scheme. This scheme could trace its genealogy back to the *Loan (Farmers' Debt Adjustment) Act 1935-1971*. In 1971, the first of the modern Rural Reconstruction Schemes was put in place and, although it was amended in 1976, 1985 and 1989, its structure and focus had changed only incrementally. In 1992 the scheme was again reviewed and refocused with a much stronger emphasis on productivity improvement and efficiency. The 1989 version of the scheme had provided support to farmers to

(a) overcome financial difficulties arising from causes beyond their control;
(b) improve their performance by changing the size of their farms, improving managerial and financial skills, or by adoption of improved practices and technology;
(c) make an orderly exit if, after all options have been considered, the farmers are without prospects in the rural industry. (States and Northern Territory Grants (Rural Adjustment) Act 1988, s5

The 1992 scheme shifted this focus to the provision of support to 'improved productivity, profitability and sustainability through structural adjustment and more effective management of the farm business'. Support was provided in the form of interest rate subsidies on commercial borrowings 'to enhance the productivity, sustainability and profitability' of farm businesses. Small grants were also available for skills improvement and for financial, planning and other advice (Crean 1992, p2413). The threshold eligibility criterion was the capacity of the farm business to obtain commercial finance. This approach was based on the assumption that the commercial financial market was best placed to judge whether a farm business was economically sustainable—it also avoided the problem of the Commonwealth competing with the commercial sector in the provision of support to farmers.

The innovation in the 1992 Rural Adjustment Scheme which was central to the implementation of the National Drought Policy was the introduction of 'exceptional circumstances' provisions into the Act. These provisions were intended to enable government 'to respond quickly and appropriately to severe downturns without undermining the direction and purpose of the [Rural Adjustment] scheme as a whole' (Crean 1992, p2413). The exceptional circumstances provisions were triggered by Ministerial declaration on the basis of a recommendation from a Rural Adjustment Scheme Advisory Council that 'exceptional circumstances exist in relation to the farm sector' (*Rural Adjustment Act 1992, s8(d)*). Once such a declaration had been made, enhanced interest rate subsidies became available, rising from a maximum of 50% to a maximum of 100% of the costs of commercial finance. The Rural Adjustment Scheme was therefore intended to support the National Drought Policy in two ways: by providing farmers with support to improve their management skills for dealing with 'normal' drought and to provide extra assistance in the event of exceptional circumstances.

In addition to the Rural Adjustment Scheme, the Commonwealth government established the Farm Household Support Scheme which was aimed at encouraging unviable farmers to leave the industry. The scheme provided income support for farmers on a time-limited basis with a portion of that support to be repayable at commercial rates of interest under prescribed circumstances. The scheme was linked to exit provisions within the Rural Adjustment Scheme that provided a financial incentive in the form of a grant to farmers who left the farm. These re-establishment grants had been part of the Rural Adjustment Scheme and its predecessors since 1971 and, in spite of a series of reports to government questioning their efficacy (Industry Commission 1996; McColl *et al* 1997; O'Neil *et al* 2000), remain as part of the Australian government's rural policy approach. At the time of writing, these grants were scheduled to be wound up in June 2004 (Truss 2003a).

The policy thinking was that the combination of the Rural Adjustment Scheme and the Farm Household Support scheme would provide the assistance needs of all farmers. To assist farmers further in accumulating financial reserves to improve risk management and cope through drought, the Commonwealth's Income Equalisation Scheme was enhanced and the Farm Management Bonds Scheme was set up. Both of these mechanisms provided favourable tax treatment to encourage farmers to put aside financial surpluses in the good years, to be drawn down in the bad years.

The National Drought Policy was consistent with the recommendations of the Drought Policy Review Task Force and the Senate Inquiry report, and thus attracted broadly bipartisan support (see for example Lloyd 1992, p3061). However, the timing of the scheme's introduction could not have been worse for a policy based on preparation for drought and risk management. The scheme came into effect on 1 January 1993 and by this time parts of Queensland and New South Wales had been experiencing worsening drought conditions since 1991. By 1993, the drought was becoming increasingly widespread and was to become arguably the worst drought of the twentieth century. In risk management terms, the timing was particularly bad. The drought followed several years of historically high interest rates and low commodity prices—even prior to the onset of drought there had been a sense of crisis in rural Australia.

4. The 1990s drought

By the end of 1993 recourse to the exceptional circumstances provisions of the Rural Adjustment Scheme was anything but exceptional. There were increasing calls for exceptional circumstances declarations and the first major shortcoming of the National Drought Policy was exposed—the lack of a definition in either the legislation or accompanying material of the pivotal term 'exceptional circumstances'. A rough rule of thumb of two years of drought declaration out of three was adopted; however, the declaration of drought remained a state responsibility and the rigour applied to these declarations varied from state to state. In September 1994 the Ministerial Council agreed to 'adopt a coordinated national approach to drought declarations' and also 'that officials work urgently to draw up a uniform system of drought declarations, sensitive to regional, as well as State conditions' (ARMCANZ 1994b, p4). The following month the Council further agreed on 'a harmonised system for considering future drought declarations' based on a common set of core criteria (ARMCANZ 1994a, p3). The development of these criteria and the opportunities and difficulties associated with achieving such a science-based approach to drought declarations are discussed in more detail in Chapter 7.

As the drought worsened in 1994, a further shortcoming in the policy was revealed—the apparent inadequacy of the policy response to the welfare problems generated by the drought. The Rural Adjustment Scheme/Farm Household Support Scheme combination proved not to cover all those in need as had been anticipated. Farmers who were not eligible for support through the Rural Adjustment Scheme were not receiving any support as a result of the exceptional circumstances declarations and the only way to do so was to sign on to the Farm Household Support Scheme which implied either increased debt or exit from farming. This choice was not appealing to farmers facing hardship as a result of the drought and throughout 1994 the Commonwealth Minister received a series of delegations from private charities, welfare providers and rural counsellors reporting on a growing welfare problem in drought-affected areas of rural Australia.

In mid 1994, a number of media organizations established the Farm Hand Appeal seeking public donations to provide relief to drought-affected farmers. The bipartisanship that had been evident at the establishment of the National Drought Policy disappeared and the Opposition parties raised the temperature of the debate in Parliament, attacking the government for its lack of response to the welfare problem:

> In 1993 as Minister for Primary Industries and Energy, Simon Crean had faced fewer than ten questions without notice on rural adjustment, drought or the rural downturn. In 1994 the new Minister, Senator Bob Collins responded to an increasing number of questions on drought-related issues with 35 being asked in the parliamentary sitting period between 23 August and 8 December 1994 alone. Increasingly emotive and colourful language was employed by Opposition members with references to the 'sheer human misery that has been created by what is possibly the worst drought this century' and suggestions that 'drought

is probably the most important issue that this parliament can currently address'. (Botterill 2003d, p68)

In September 1994 the then Prime Minister visited one of the worst affected parts of Queensland and in a stirring speech he 'extended his election promise not to leave the unemployed behind, to the country people of the nation' (Wahlquist 2003, p80). In early October the Government announced the establishment of the Drought Relief Payment. The payment marked a departure from the strong structural adjustment focus of the Rural Adjustment Scheme by making welfare payments to *all* farmers in exceptional circumstances areas, not just those with a long-term future in the sector. The payment was taken up very quickly, and 10,500 farm families were accessing the scheme by May 1995 (Collins 1995, p512).

Although in welfare terms, the Drought Relief Payment was a major success, it complicated the implementation of the National Drought Policy by making an exceptional circumstances declaration considerably more desirable and increasing pressure on policy makers to develop an objective and equitable definition of exceptional circumstances.

5. Between major droughts 1996-2001

In 1996, there was a change of government at Commonwealth level and there was a series of reviews of the rural policy approach. A review of the National Drought Policy had been initiated in 1995 and it reported in early 1997. The Rural Adjustment and Farm Household Support Schemes were also reviewed. The National Drought Policy review was set up to 'cover all measures identified under the National Drought Policy, including an examination of: welfare related assistance, financial and institutional impediments, taxation arrangements available to the farming sector and an assessment of assistance measures and delivery mechanisms' (ARMCANZ 1995, p86). The review proposed a revised set of objectives for the National Drought Policy:

- *to encourage primary producers and other sections of rural Australia to adopt self-reliant approaches to managing for climate variability*
- *to maintain and protect Australia's agricultural and environmental resource base during periods of extreme climatic stress*
- *to ensure farm families are provided with adequate welfare support commensurate with that available to other Australians*
- *to ensure that the elements of the National Drought Policy do not impede structural adjustment, and*
- *to have a high level of awareness and understanding of drought and drought policy.* (Drought Policy Task Force 1997, p.6—italics in original)

In February 1997, the Ministerial Council 'accepted the need to integrate the approaches to risk management, adjustment and welfare' and also that 'business support needed to be reoriented away from relief measures, including interest rate subsidies, and towards preparedness measures' (ARMCANZ 1997, p19).

The review of the Rural Adjustment Scheme, which had been conducted concurrently with the review of drought policy, concluded that the scheme was 'not appropriate to the adjustment needs of Australian agriculture in either today's business environment or that expected in the next century' (McColl *et al* 1997, p ix). In 1997 the Commonwealth Minister announced that the scheme was to be wound up and replaced with a package of measures called *Agriculture—Advancing Australia* (the 'AAA' package). The Minister identified four key objectives of the package:

- to help individual farm businesses profit from change;
- to provide positive incentives for ongoing farm adjustment;
- to encourage social and economic development in rural areas; and
- to ensure the farm sector had access to an adequate welfare safety net (Anderson 1997b).

The exceptional circumstances provisions of the Rural Adjustment Scheme were set up as a standalone Exceptional Circumstances Scheme and the Drought Relief Payment became the Exceptional Circumstances Relief Payment, extending the payment to farmers affected by any form of exceptional circumstance, not just drought. The Farm Household Support Scheme was abolished and replaced with the Farm Family Restart Program (later renamed Farm Help) and the re-establishment scheme was retained as part of this program.

In March 1999, Ministers again considered the exceptional circumstances policy approach. A new set of exceptional circumstances guidelines was agreed on, with the criteria for an exceptional circumstances declaration to be that:

(i)   The event must be *rare* and *severe*.
(ii)   The effects of the event must result in a *severe* downturn in farm income over a prolonged period.
(iii)   The event must not be predictable or part of a process of structural adjustment. (ARMCANZ 1999a, p59)

Where the earlier indicators had specified meteorological conditions to be the threshold issue for an EC declaration, the 1999 decision was that 'the key indicator is a severe income downturn, which should be tied to a specific rare and severe event, beyond normal risk management strategies employed by responsible farmers.' The severe downturn should be 'for a prolonged period and of a significant scale'. 'Rare' was taken to be an event which occurs 'on average *once in every 20 to 25 years*' (ARMCANZ 1999a, p 63). The exceptional circumstances issue was again on the agenda for the Ministerial Council in August 1999 with Ministers agreeing that

(a)   future Exception [sic] Circumstances (EC) applications will not be submitted to the Commonwealth until the cases can be fully documented and the State or Territory Minister (or peak industry body) is reasonably confident that the case fully meets EC guidelines; and

(b)   members should actively pursue measures to increase farmers' ability to manage their own risk. (ARMCANZ 1999b, p4)

The first of these points reflected the political problem associated with a scheme in which the States nominated regions for declaration but the Commonwealth provided the lion's share of funding. There was no incentive for States to scrutinise applications closely and attract any resulting criticism from farm groups for rejecting their case. It was far easier politically to let the Commonwealth reject unsubstantiated applications for support.

Following a workshop of officials in October 2000, a number of key points for consideration were put to Ministers. It was suggested that exceptional circumstances (EC) policy should:

a)   provide assistance for a single rare climatic event (with allowance for compounding factors);
b)   cover events outside the range of "normal" business/enterprise risk management and be confined to people involved in a farm business;
c)   not impede longer term, "normal adjustment";
d)   support the development of "normal" risk management and sustainable resource use, with this in mind it was agreed that farmers receiving EC assistance be required to undertake business planning; and
e)   provide for improved integration of State and Federal support measures that encourage farmers to manage their own risk (ARMCANZ 2001a, p 44).

6. The 2001-03 Drought

From 2001, areas of Australia began once again to experience deteriorating drought conditions and by October 2002, the Bureau of Meteorology was describing the seven-month period from April to October 2002 as "remarkable for its spatial extent of rainfall deficiencies and average level of "dryness" with "nearly half of the country recording their driest October on record with much of NSW registering record low totals" (BOM 2002). Once again, this meant that Ministers were considering exceptional circumstances policy in a highly-charged political environment.

An ongoing criticism of the exceptional circumstances program since it began in 1992 has been that declarations have been based on administratively defined geographical areas. This has resulted in both perceived and actual inequities between farmers in objectively similar circumstances who fall on different sides of the line. This so-called "lines on maps" problem was addressed by Ministers in August 2001 when they agreed that

> Farmers outside the defined zone, but who are in reasonable proximity and can also demonstrate that they are affected by the same exceptional events, will be eligible to make application under the same terms and conditions as those within the defined zone. (ARMCANZ 2001b, p33)

This blurring of the lines reduced the criticism of this aspect of the EC program; however, it has not addressed the inequity. A further criticism of the implementation of the EC program related to the length of time it was taking for an application for EC support to be considered and thus for support to be made available. In September 2002, the Commonwealth Minister announced that income support payments would be made available to farmers on the basis of a *prima facie* case that they qualified for support so that they did not have to wait until their application had been assessed fully. Under this arrangement unsuccessful applicants would receive up to six months of welfare support at the equivalent of the unemployment benefit while successful applications resulted in income support payments for two years along with business support, the latter being limited to viable farmers.

In October 2003, amid signs that the drought may be coming to an end (BOM 2003), the Ministerial Council announced that a national roundtable would be convened in 2004 to "discuss how future drought assistance can be made more efficient and effective" (Truss 2003b). The Roundtable will follow consultations by an independent panel with stakeholders and a range of options will be developed for consideration.

7. Drought policy challenges and the way forward

Like other drought policy reviews before it, the 2004 roundtable faces a number of challenges in arriving at an effective, equitable and sustainable drought policy. As the history of the National Drought Policy has shown, there is considerable pressure on Ministers to act during drought to adapt or ease policy settings and this has resulted in *ad hoc* policy change which is arguably inconsistent with the underlying principles of the National Drought Policy. Those principles of self-reliance and risk management remain largely uncontested by members of the policy community. For example, the NSW Farmers' Association's drought discussion paper of October 2003 focuses on drought preparedness and a recognition that 'Preparing a property for drought and developing a plan about how to manage through drought is an essential part of normal operations for primary producers' (NSW Farmers' Association 2003, p6). For policy makers, translating these agreed principles into a workable and equitable drought response means addressing a series of related problems.

7.1 THE PROBLEM OF DEFINITION AND POLITICISATION

The question of developing a science-driven definition of exceptional circumstances and the politicisation of the drought declaration process is addressed in more detail in Chapter 7. For the purposes of this discussion, it should be noted that a definition needs to be not only scientifically sound but also must be seen to be fair in order to reduce the likelihood of lobbying for change once the policy is implemented. An effective drought policy needs to minimise the opportunities for politicisation while recognising that, in a liberal democracy, politicians are responsive to public opinion and lobbying.

7.2 THE DECLARATION PROCESS

The declaration process can be constructed in a number of ways. In 1999, Ministers stated clearly that 'the EC program has not been designed to provide support for individuals, rather it has a regional focus' (ARMCANZ 1999b, p4). This approach has been problematic as the regions on which drought declarations have been based have been administrative areas and have not been related to topographic or biophysical characteristics.   For example, in the southeast corner of the Bourke-Brewarrina administrative area in New South Wales there is a narrow 'finger' of less than 40 kilometres long and barely 15 kilometres wide which extends south into the Nyngan administrative area. It is clear that an exceptional circumstances declaration which includes Bourke-Brewarrina but excludes Nyngan will create equity problems. Declaration on an individual farm basis also raises issues of equity, particularly relating to the difficulties of distinguishing an EC arising from agronomic drought as opposed to management-induced problems. The experience of the individual property declarations in Queensland under the natural disaster relief arrangements also contains lessons, particularly if there is any scope for discretion in the declaration process.

It could be argued, however, that a risk management approach to drought requires an environment in which individual farmers have access to management tools such as the highly successful Farm Management Deposits Scheme. A scheme of this type is focused on the individual and not the region and could be supplemented by similar schemes such as government-backed income related loans (Botterill and Chapman 2002). A policy based on providing farm managers with the tools to prepare for drought could remove the need for declarations altogether, thereby removing a number of political pressure points from the process (see Chapter 7).

7.3 FARM WELFARE

A big obstacle to the ending of the system of drought declarations is their central role in the delivery of welfare support to farmers during exceptional circumstances. The welfare component of the current exceptional circumstances program is highly valued by farmers, particularly smaller farmers (Gaynor, pers comm), so any change to the program will need to ensure that the welfare needs of farmers in drought are met.

In general, delivering welfare support to farmers in need has proved difficult. The general welfare safety net in Australia is targeted at wage and salary earners or categories of pensioners such as carers, retirees or those with a disability. It does not respond well to the needs of the self-employed or those who are income-poor but asset-rich. This has been particularly the case since the introduction in the mid 1980s of assets tests for welfare payments. The return on farm land is low in comparison with other investments and therefore many farmers who are not income-rich own significant assets. When incomes fall for whatever reason, piecemeal liquidation of the asset is not practical, particularly as this could seriously undermine the ongoing viability of the farm operation. Many farmers are addressing low incomes through off-farm activities, either their own or those of other family members. When those family members are unable to obtain suitable employment they find themselves excluded from the unemployment benefit, often due to the value of the land of which they are a joint owner.

Since 1989 a series of schemes has been implemented by the Commonwealth government aimed at meeting the welfare needs of farm families: from 'Part C' of the 1988 version of the Rural Adjustment Scheme through to the Farm Help program in place at the time of writing. All of these programs have been hindered in their effectiveness by the desire of governments to meet two apparently conflicting objectives: the delivery of welfare support to those in need and the achievement of structural adjustment in the farm sector. This tension is a particularly difficult issue for policy makers. There are strong moral and equity arguments for the provision of welfare relief to members of the community unable to meet basic day-to-day living expenses and the delivery of this type of support is therefore politically salient. However, income support of this nature can mask structural adjustment problems within the farm community and can trap farmers in a situation which generates social, economic and environmental problems. Due to the integrated nature of many family farms, the provision of welfare assistance runs the risk of providing a *de facto* subsidy to an otherwise unviable farm business. This concern has resulted in programs over time which have either been time limited, repayable or linked to exit from the industry (Botterill 2003a). The Exceptional Circumstances Relief payment provides assistance to a group in the community which otherwise has difficulty obtaining government income support. If drought declarations were to end completely, the issue of welfare support for farmers would need to be revisited.

7.4 POLICY CHANGE DURING DROUGHT

As outlined above, the National Drought Policy was introduced as Australia was entering a severe drought and was amended both during the 1990s and again in the 2001-03 drought. Amending drought policy during a drought event means that the policy process is taking place against a background of hardship and political point scoring. Such an environment is not conducive to a thoughtful or rational policy process. Although the existence of drought can focus policy attention on drought response measures, it would be more desirable for the issue to be debated when conditions are better. It is hoped that this will be the case for the 2004 review and that a sustainable policy can be developed which attracts sufficient consensus that there is not the level of pressure for *ad hoc* changes to policy settings during the next severe drought.

7.5 COST AND EQUITY

Current drought policy settings are expensive. In October 2003, the Commonwealth Minister announced that expenditure on the 2001-03 drought had exceeded $A1 billion. Spending of this magnitude raises questions of equity in terms of both the distribution of funds to farmers and between farmers and other groups in the community. As has been argued elsewhere (Botterill and Chapman 2002), the current funding arrangements are regressive as the farmers in receipt of drought support are in many cases wealthier in a lifetime sense than the average taxpayer contributing to that relief. Funding options which involve a loan component could address this issue, possibly along the lines of the low interest loans offered under the natural disaster relief arrangements. As pressures continue on governments to run surplus budgets and retain funding for the provision of other public goods, substantial expenditure on drought relief could come under closer scrutiny. Some commentators are already questioning government largesse during drought, observing, for example, that 'Farming must be the only for-profit industry in the country that passes round the hat when profits slip. If any city businesses tried that, we'd laugh them to scorn' (Gittins 2002).

7.6 COMPLEXITY AND TRADE OFFS

The impact of drought is not limited to farm businesses and the welfare of farm families. It also affects non-farm businesses, rural communities and the environment. The policy response under the National Drought Policy was initially focused on the farm business with policy measures aimed at sustaining the farm economically. In recent years, the policy has moved more towards addressing the welfare needs of farmers affected by drought. Attempts have been made to address the environmental impact of drought through funding for drought-related Landcare activities.

As noted above, a policy instrument aimed at delivering welfare relief can inadvertently undermine structural adjustment by keeping on the land farm managers who otherwise would leave. Similarly, economic instruments such as transaction-based subsidies, for example fodder subsidies, can slow the rate at which farmers destock their properties with the onset of drought, resulting in environmental damage. Transaction-based subsidies were to have been phased out by State governments under the National Drought Policy in recognition of their potential adverse impact on the environment, as well as the possibilities they offered for rorting, however, they are still in place. One of the biggest challenges facing policy makers is addressing the trade-offs between conflicting values such as these. There is no single objectively correct drought policy for Australia—any combination of policy measures will privilege one value over the others. The difficulty in delivering a policy which attracts a bipartisan consensus is in balancing competing values in a way which is perceived by stakeholders to be fair (for further discussion of this trade-off in the context of Australia's drought policy see Botterill 2003c).

8. Concluding remarks

Australia has made considerable progress in the development of a realistic response to drought. Since 1992, the policy community has proceeded on the basis that drought is a normal feature of the Australian environment and not, as one commentator described it, 'some utterly unexpected act of bastardry on the part of the Deity' (Gittins 2002). This has shifted the focus from a disaster response to looking at the best means for farmers and government to prepare for and manage the recovery from drought events. Unfortunately the caveat to that approach has been to recognise that occasionally severe droughts occur which are beyond the capacity of the best manager's risk management skills. The frequency with which the exceptional circumstances provisions have been invoked since they were first developed means that the focus has not been on managing droughts but rather on making a case that government intervention is warranted. The exceptional circumstances process suggests that in practical terms we have not moved as far from a disaster response as the letter of the National Drought Policy would suggest. The 2004 policy review provides an ideal opportunity for a serious rethink of the policy instruments on offer before, during and after drought. The objective should be to assist farmers in managing climate uncertainty in a manner which depoliticises the process as much as possible and delivers equity and predictability to the environment within which Australian farmers operate.

CHAPTER 5: MANAGING RISK?: SOCIAL POLICY RESPONSES IN TIME OF DROUGHT

DANIELA STEHLIK
Alcoa Research Centre for Stronger Communities, Division of Humanities, Curtin University of Technology, Building 100, GPO Box U1987, Perth, WA 6845, Australia

> It should be a natural disaster, you can't get any more natural. It is like someone up there turned the tap off (Female grazier, CQ).

> A drought is only a drought if it is out of your control ... if it's total out of your control, [then] yes, there should be some assistance (Female wheat/sheep farmer, NSW).

1. Introduction

Drought is integral to the Australian identity and lived experience. Our historical and cultural understandings as a nation about what we mean by a natural disaster have always included drought. It would be uncommon to hear ordinary people—even those living in cities—talking about drought as something that could be *managed*. Yet, despite this broader societal acceptance of drought as disaster, the fact is that since the late 1980s, Australian policies have officially *not* recognised drought as a natural disaster, thereby exposing a policy paradox.

This paradox of *non*-disaster also enables a re-conceptualisation of the way in which drought in Australia has now been framed as a risk management issue, one which 'fits' with a self-reliance ideology (see Higgins 2001). Drought is now no longer viewed as an external force, one that cannot be controlled. It is, instead, viewed—certainly in policy and government management—as an internal farm matter, one that should be anticipated for, then controlled and managed, and therefore is something that becomes the responsibility of the individual farmer and the whole family to deal with. This chapter suggests that if drought impact is not viewed as a collective matter but rather one left to individuals to resolve, it thereby follows that a building of community capacity as a response becomes more difficult.

In this sense, policies responding to drought reflect the overarching neo-liberal philosophy that has underpinned Australian governments (and those in western nations generally) of all political persuasions since the mid 1970s. Responses to drought in Australia just took a little longer to get caught up in this market driven perspective, but since the late 1980s, as this chapter shows, the transition from disaster to managed risk has challenged the way in which those affected by drought have been dealt with, as well as challenging their own sense of identity in the face of ongoing drought events. The

65

L.C. Botterill and D.A. Wilhite (eds.), From Disaster Response to Risk Management, 65–83.
© 2005 *Springer. Printed in the Netherlands.*

impacts of the changes to policy in the 1980s were keenly felt in the drought of the subsequent decade.

This chapter takes a historical perspective by tracking the changes in social policy responses to the Australian drought of the 1990s and suggests that the 'move' to self-reliance created a dichotomy between policy development (urban) and policy implementation (rural) and a confusion as to service delivery responsibilities particularly in the light of the public/private provider split. Evidence gathered during the 1990s drought from farm families in New South Wales and Queensland explores the transition from 'disaster' to 'managed risk' in more detail. The chapter begins by establishing the policy and environmental context in which the move from disaster to managed risk was promulgated. It then briefly describes the research project undertaken in the 1990s, from which evidence as to the impact of changes in social policies was drawn. These impacts are then discussed, drawing on the comments from our respondents, as well as from statistical data. The paper concludes with an analysis of the long-term implications of policy decisions made in the 1990s for the current drought and future droughts.

2. A political context

Pinker (1973) writes that the 'study of social welfare is a study of human nature in a political context' (p211). Nowhere is this in clearer evidence than in an analysis of the social policy responses to drought in Australia in the past two decades. The changes to social policy towards a neo-liberalist framework commenced in Australia in the early 1980s with a series of Federal government reviews of community services. These marked the move to a market model, one more corporatised in its practice. This change also marked a decided shift in the relationships between the three tiers of Australian government—Federal, State and local. In the 1990s, Federal governments adopted an increasingly 'hands off' approach to service delivery, while retaining their benchmarking and funding role. Increasingly, in rural and remote communities, Federal representation was withdrawn, and State governments were left with the responsibility for a 'presence', particularly in the large regional centres. This transition accelerated after the Federal election in March 1996, when the conservative Liberal-National Party Coalition was elected, despite one of its members being a party that traditionally represented the 'bush' constituency.

Pinker suggested that while social welfare in a western industrialised context derived essentially from 'collectivist ideologies', the tendency to a neo-liberal, market dominated view of self-help and individualism stresses a difference in the conceptualisation of citizenship. For some, he writes, 'citizenship is enhanced and extended by the existence and use of social services. For others, citizenship is debased by reliance upon such aid' (Pinker 1973, p201).

For the citizens of rural Australia, the paradigm shift to 'risk' did indeed touch on issues of identity, community, citizenship and their place in society. The legacy of this transition, as Hancock suggests, 'privileges the market over the social, and [puts] increasing pressure on individuals and families to cope with new and emerging patterns of risk' (2002, p132). Pusey, a strident critic of the economic rationalist approach to social welfare, argues that it 'is a kind of laboratory test of what are in effect metatheories of society' (Pusey 1991, p240). As one commentator at the time noted, the government 'cannot go much beyond the free market nostrums, despite the enormous pressure from the bush to do something different' (Hamilton 1996, p7).

A central component of these 'free market nostrums' was the ongoing and rapid separation of the purchaser/provider split—or, in other words, the privatisation of services previously within government responsibility[1]. This has impacted on rural/regional Australia both through uneven distribution, as some areas have lost services that have not been replaced, even in a private capacity, and in the loss of 'local expertise, knowledge and networks' (Alston 2002, p98). In this way the opportunity to build community capacity at the time of the greatest crisis experienced by many rural families and communities was lost as the focus was *not* on 'innovation and social capital building'. Alston continues:

> Despite high levels of disadvantage, poverty and reduced quality of life, many rural people have been seriously disadvantaged by devolution, further exacerbating social exclusion and alienation. Under such conditions, social capital at local level is being seriously thwarted (2002, p100).

The trend towards these reforms created serious challenges for rural communities, including the potential for inequitable resource allocation; unrealistic expectations of community management within small communities, where 'everyone does everything' usually on a voluntary basis; and a tendency to focus on outputs, rather than outcomes.

Although beyond the scope of this chapter, it is nevertheless instructive to note that social policies in North America are similarly affected by policy decisions that make assumptions about rural populations. Christenson and Flora (1991) also make the point (highly relevant to Australia), that 'national policies relative to rural populations in the rest of the world influence the social and economic conditions of our country' (p334).

The intersection between the market ideologies of neo-liberalism, globalisation and the extended drought of the 1990s created a policy environment which, as Schram describes, focuses on 'solutions' rather than an agreed definition of the problem. Such problems, he suggests, are

---

[1] One example is the establishment of Centrelink, a 'one stop shop' which offers access to all Federal human service agencies usually located in a regional centre. Rural telecentres have been established for those living outside regional environments. Centrelink is not considered a mainstream public agency, and replaced Commonwealth Employment Service (CES) in many smaller communities. Another example, not in human services, is that of the ongoing sale of the telecommunications company— Telstra.

Defined in a particular way so as to justify treating them according to another policy approach … in this sense, policies create problems: each policy creates its own understanding of the problem in a way that justifies a particular approach to attacking the problem (1995, p125).

3. The problem: when is a drought a crisis?

As the science around climate variability has become more accessible, the ties between disaster and climate have been loosened. This was recognised in 1992 by the then Goss Labor Government in Queensland, in a policy paper which suggested that drought was different from cyclones and fire as 'it has no obvious physical presence' (Queensland Government 1992, p7); thus the actual physicality of drought is rejected, an irony not lost on those who have experienced drought conditions. In a publication dated 1994, supported by the Queensland Department of Primary Industries, a 'disaster' drought was defined as once every 18 years (Daly 1994, p91). Another suggested once every 20-25 years (White *et al* 1995, p256). In this way, the droughts of the 1980s and 1990s, coming so close together, being so widespread and impacting for such long periods (in some cases over 6 years), definitely did not 'fit' such definitions.

It was the federal Drought Policy Review Task Force established by the Hawke Labor government in 1989 which confirmed the paradigm shift (see Chapter 4 for more details). In the shift from disaster to self-reliance, it re-framed drought from an external event that could not be predicted, to one representing 'a prolonged failure or inability of producers to respond to those deteriorating conditions.' The Task Force report became quite clear about where the responsibility for drought lay. It wrote

Drought is…a *relative concept* that reflects the fact that the current agricultural production is out of equilibrium with prevailing seasonal conditions. Managing for drought, then, is about managing for the risks involved in carrying out agricultural business in a variable climate (DPRTF 1990, p7—my italics).

Simmons suggests that the reason for this shift was 'based on perceptions that drought was a relatively slow phenomenon and that the general occurrence of drought could be anticipated and thus people could prepare for it' (1993, p445). This was compatible with neo-liberal views and the increasing trend towards selectivity in social welfare policies (Jamrozik 1983). However, it resulted in drought no longer being seen as an 'Act of God' but 'normal risk', responsibility for which was 'shifted solely to the individual farmer who was expected to conduct him or herself in a financial prudent manner' much in the same way as a share broker would work on 'market fluctuations' (Higgins 2001, p124). In addition, the statement about 'carrying out agricultural business' places the impact of drought in the economic, rather than the social or emotional, sphere, thus avoiding the recognised fact that for many, farming is a way of life, rather than an economic undertaking. As Hamilton argues,

...economic incentives appear to be subordinate to life-style and cultural factors and ... this helps to explain the tenacity with which farmers remain on the land in the face of sometimes overwhelming financial difficulties (Hamilton 1996, p7).

Taking the 'disaster' out of drought meant that events moved outside of the purview of Natural Disaster Relief Arrangements (NDRA), where arrangements had been in place for over thirty years, and shifted them to state and local governments as well as to individuals. At the same time, other disasters that remained 'natural', such as floods, bushfires and cyclones, remained in the NDRA framework. The Commonwealth had borne the burden of the concessional loans, freight concessions and subsidies which enabled farmers to carry on in times of crisis through the NDRA, while the states also shared some fiscal responsibilities. In a practical sense, such arrangements were largely processed through agricultural departments (Kerin 1987), thus supporting a production (support) model, rather than a welfare (dependence) model. This is a crucial point, as we shall see, because for those families receiving such support at that time there appeared to be no perceived stigma associated with it.

By the time the Industries Assistance Commission (IAC) reviewed the Rural Adjustment Scheme (RAS) in 1983 (see Bryant 1992), it heard criticism of abuses of the NDRA, and from economists who were suggesting 'serious flaws' in transaction-based policies such as fodder and freight subsidies (Freebairn 1983). Two of these included: appearing to favour sheep and cattle graziers, rather than all agricultural industries; and deterring some from preparing for the future (that is, to quit the farm), by appearing to just 'rely' on relief when it was needed. This was also the time of the growth of conservation consciousness, and a broader consensus was emerging that soil and water conservation measures were not being addressed adequately in the current financial arrangements. For example, Bryant suggests that incentives for 'sound land use practices [were] provided to those farmers who financially did not need them' (1992, p168).

However, despite the IAC reporting that 'the farming community did not appear interested to establish the need for any significant additions to the existing measure for drought assistance' (IAC 1983, p2), nevertheless it did take up the submissions of many witnesses other than the farming community, including the banks, which appeared to confirm that drought was an opportunity for some to make profit from government subsidies on NDRA arrangements. As quoted in Daly (1994, p94) it was felt that some 'property owners lived off properties in above-average seasons and off government assistance in below-average seasons'.

4. Managing risk through welfare?

It was not surprising then to find that the welfare component of drought assistance became the key, in subsequent years, in managing the risk. The policy discourse talks of 'hardship', 'welfare' and 'support'. In this sense, such discourse identifies a disadvantage experienced by some, which the policy is designed to alleviate in some way. Central to the 'hardship' issue was the argument that existing policies tended to encourage non-sustainable farming. At a more broad, philosophical level, such a debate can be seen as one focussing on the 'deserving' and 'undeserving'—a central component of the universalist versus selectivist approach to social welfare (Colebatch 2002). In other words, those who farmed sustainably should be supported, those who did not should be encouraged to leave.

This important aspect of the developing 'non-disaster' approach hinged on policy language about 'exceptional circumstances'. When was a drought exceptional? Schram (Schram 1995, p xxiv) argues that the discourse of welfare policy has 'symbolic consequences in reinforcing prevailing understandings' and the language (discourse) of exceptional circumstances provides us with a classic example of how policy shapes the actions of those it is meant to support and how it has unintended consequences as it does so. (For a detailed analysis of exceptional circumstances see Chapters 4 and 7 in this volume and Botterill 2003d). In a time of turbulence, state governments developed alternative approaches. For example, the Queensland government—already experiencing pressure from a rural sector in drought for at least three years—argued for an inclusion of exceptional circumstances in the Rural Adjustment Scheme on the grounds of productivity (Higgins 2001, p143).

By 1994, the eastern seaboard of Australia was gripped by drought and it was during this time that the transition from disaster to risk (in terms of policy) was completed when the then Prime Minister, Paul Keating, visited Central Queensland, late that year, and talked of 'managed risk' and 'way of life' rather than disaster. As the drought continued, policies began to be realigned to the reality, and in the same year the debate shifted to an expansion of the concept to include the welfare component through the Drought Relief Payment. This should not be seen as a return to drought as 'disaster' but rather as an opportunity to realign policies to continue to place pressure on individuals in terms of their own risk management. Gow (1997, p 278) suggests that this was a political decision which enabled a 'pigeonholing' of a potential political issue.

Such assistance packages were heavily weighed with responsibilities, including decision making regarding financial counselling. Such assistance packages were also reliant on 'declarations' and assistance could not be provided unless the declaration was made for a district, which was the direct responsibility of particular local governments or of local boards. When the assistance came, however, it was too little too late for many. As Rowlands (2000/01) points out, in social policy, the 'elapsed time between announcement and implementation can take [up to] two years' (p79), a fact confirmed by respondents to the study. The severity of the 1990s drought also tested the capacity of the three tiers of government, a complex relationship in 'normal' times, to develop and deliver social policy immediately (Commonwealth Department of Human Services and Health 1994, p115). They strained under the pressure to collaborate with each other during the crisis. The lack of integrative relationships between agricultural departments and social security departments during the crisis was also highlighted as a challenge to the timely delivery of services (Stehlik and Lawrence 1999).

It was very soon after the election of the conservative Liberal-National Party Coalition government in March 1996 that press releases began to highlight the paradigm shift. The first discourse that began to emerge talked of the 'two Australias'—the urban and the rural. This analysed the election results in terms of a rural voter back-lash against urban Australia—particularly the Sydney/Melbourne/Canberra triangle, and then talked of a 'decoupling' of the big cities and regions of Australia (Rodgers 1996, p25). There remained a basic tension in government rhetoric and policy, as on the one hand while the argument for regional representation and access to services on an equal basis to those in urban centres was central, there was also a continued diminution of services based on a per capita rather than needs basis and argued from a market perspective, so that if the rural population declines, the services were withdrawn (Stehlik *et al* 1996). Drought very much became a political issue and a testing ground for neo-liberalism (Gow 1997).

The emerging 'two Australias' was, unfortunately, also supported in the subsequent transition of social policy from agricultural subsidies (production) to welfare (dependence) continued. Less than six months after the March election the headlines read: 'Welfare to replace farm drought aid'. The release quotes the Minister for Primary Industries—John Anderson (also Deputy Prime Minister and Leader of the National Party), who, while still relying on some of the older paradigm, argued that:

> ...using the social security system to pay farmers affected badly by drought or other natural catastrophes was a fairer, more effective approach (McKenzie 1996, p11).

Here the unintended, yet crucial, impact becomes one associated with stigma. While policy was being delivered by agricultural agencies and badged as 'disaster relief', those receiving the benefits were still focused on the 'production' aspect of support. As the transition to social security took hold, and people became 'caught up' in the so-called 'safety net' of welfare, their own sense of identity as a worthwhile contributor to the national enterprise was challenged. In addition, for many, their own, previously discriminatory attitudes towards so-called 'dole bludgers' or 'welfare cheats' were also confronted. Pinker's suggestion of 'debasement' (Pinker 1973, p201) was borne out as many in Australia's rural heartland had to join the unemployed in seeking assistance from the social welfare system, thus contributing to their sense of alienation. As the next section describes, the fairness and equity of such policy responses to drought did not reflect in people's every day experiences.

4.1 'MOMENTS OF CRISIS': LIVING THE POLICY IMPACTS

The 1990s was a decade of increased demographic changes in rural and remote Australia. The movement of younger people seeking work opportunities in coastal cities resulted in an ageing-in-place population (Higgins and Stehlik 1999; Stehlik 1999; Stehlik and Lawrence 1996). The impact of the drought, falling commodity prices and a shift to information technologies in agri-business meant a diminution in on-farm casual labour as well as a growth in off-farm paid labour by women (Jennings and Stehlik 2000). This can be seen both as a cultural differentiation and a change in structural relationships within groups in rural Australia, as the traditional hierarchy of graziers, farmers and townspeople has been redefined, with a growth in some parts of the country of in-migration of those seeking retirement, or employment in mining, tourism or the service sectors. In 2002, for example, there was a reduction of 100,000 jobs in rural Australia (Baldock 2003). Between 1996-2000, the number of Australian farms fell by 10%, with large farms (over A$500,000 income per annum.) increasing by 32%, and smaller farms (less than A$50,000 income per annum) falling by 18% (Pritchard 2002). Between 1983 and 1993 it was calculated that the return on assets of family farms was a 'mere 0.15 per cent' (Hamilton 1996, p2).

Such a transition of population has a number of impacts, not all at the economic level. Those who specifically undertake agricultural pursuits may view such in-migration with ambivalence, and as Gray (1991) has shown, conservative community leaders may impose what they believe to be 'the common interest' for the town or district, thus effectively marginalizing those with little voice to demand action. Out-migration also has its impacts, particularly on the cohesion of communities.

The impact of the drought resulted in what Marston, drawing on Fairclough, suggests can be seen as 'moments of crisis' when 'the nature of social relationships and social identities' are uncovered 'because they make conflict between parties more visible and apparent to the observer' (Marston 2000, p353). The decline in social networks, and the diminution of social cohesion as the crisis continued, exposed a dichotomy not just between urban and rural but also *within* rural communities as the much vaunted 'rural lifestyle' fast disappeared.

At the peak of the 1990s drought, research across two states funded through the Rural Industries Research and Development Corporation (RIRDC) undertook the first sociological study of its kind into the impact of drought on farm families, not only in terms of the questions asked, but also regarding timing. While many studies of 'disaster' explore a response post-event, this research allowed people to reflect on its impact *while they were still experiencing it*. In order to capture the differences in agricultural production, the research was conducted in two discrete areas—the cattle grazing region near Rockhampton in Central Queensland (CQ) and the wheat/sheep region around Balranald in western New South Wales (NSW). A major assumption underpinning the research was that men and women would experience the drought differently, and thus the project was established to capture this difference (Stehlik *et al* 2000).

The study consisted of semi-structured interviews conducted with the same questions being asked of both husband and wife in separate interviews conducted on the same day at their home; six focus group discussions—both with male and female producers and with stakeholders, such as extension officers, environmental officers, doctors, rural nurses, community development personnel and local small business owners—were conducted. In total, 103 adults (52 women and 51 men) on 56 farms in western New South Wales and central Queensland were interviewed. Their average age was in the 45-49 cohort; their average property size was 9,000 hectares and they had around 25 years of experience in farming. The project reported to the RIRDC in 1999, (Stehlik *et al* 1999) and a more detailed discussion regarding the method used can be found in that report.

This next section takes a broad view of some of the key responses, and the section following analyses the impact of the paradigm shift in more detail. Where possible, figures explaining the data are provided. In an analysis of the effect of drought on farm input it can be seen that for more than one-third of the respondents, the drought had either eliminated farm production altogether or reduced it to its lowest level ever. Of the total, 77% reported that farm production contributed to 100% of their income, thus the reduction in production had a direct and immediate impact on their financial viability and thus on capacity to maintain a quality of life. There is a marginal difference visible between New South Wales and Queensland in this regard as the following figure shows.

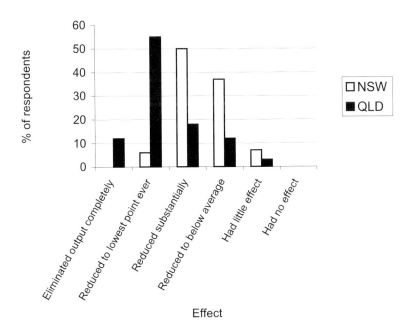

Fig. 1 Effect of drought on farm input

Apart from the crucial decision about staying on farm or selling up, selling stock was the major management decision made during the drought. In many cases, 'stock' meant breeding stock, often built up over many years. A decision to sell breeding stock was seen as a 'final solution' to the impact of drought and it was done recognising that recovering post-drought would take longer. In terms of 'risk' management, this was a major decision for the whole family and also had direct implications on the bigger question of 'staying' or 'going'. A marginal difference between the states can be seen here, which is attributed to the fact that those in NSW were grain producers as well and they sometimes had feed stocks while those in Queensland did not. The majority from both states responded that this decision reduced their income substantially or to its lowest point ever. Therefore, as the next figure details, family income can be seen as being under threat and the need to turn to human services for support becomes paramount.

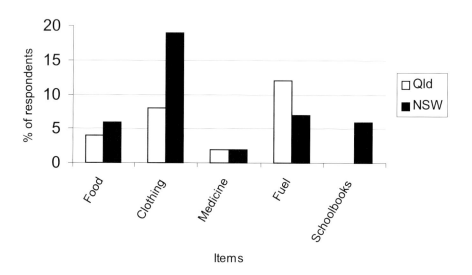

Fig. 2 Diminishment of financial resources for selected items

Decision making regarding use of diminishing income is reported in this figure. Respondents reported running out of money for essentials at least sometimes—and fuel and clothing were cited most often as items they did without. Interviews also identified that decision making regarding resources had a direct impact on the capacity of families to continue to engage in social activities in their communities. Lack of fuel meant that use of vehicles could only be justified in terms of production, not pleasure. As documented elsewhere, such crucial decision making was often gender biased, with men making the majority of decisions (see Stehlik *et al* 2000 for further discussion).

Not surprisingly, people reported that the impact of drought created personal stress and they reported other family members experiencing similar stress. This figure shows that personal stress in Queensland and NSW is relatively similar, with some respondents in NSW identifying being 'very rarely' stressed, while some in Queensland identified being 'very often' stressed. Thirty-six percent identified some effect of the drought on their health, while three percent reported 'extensive' impact. Eighteen percent reported some effect of drought on their children's health.

The RIRDC report concluded that stress has to do with 'attempting to keep the farm intact—and the animals, crops and soils in good health—while at the same time gaining sufficient income to keep the property economically viable' (Stehlik *et al* 1999, p62). Thus the combination of finding and keeping off-farm work, managing the financial obligations, and the need to maintain family relationships combined to create a stressful environment which in turn rippled out into the community as a whole. (For a broader impact, see King 1994 for a discussion on rural youth suicide.)

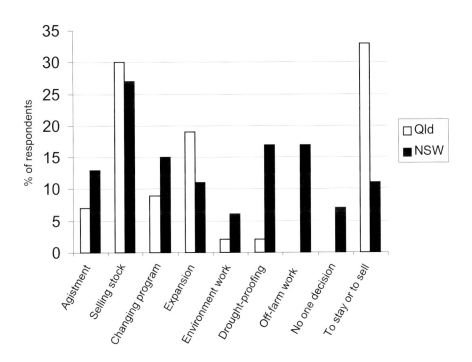

Fig. 3 Decision making as management strategy during drought

In regard to responses to stress and accessing support services, only eleven percent reported using a personal counselling service, despite seventy-three percent stating they were aware that such personal counselling was available. Ninety per cent said that they believed that some people may have been reluctant to ask for help when they needed it. Fewer men than women reported seeking counselling. Respondents also reported preferring to discuss the drought with their family members and other producers rather than with agricultural departments. One-third of respondents had not discussed drought early in its onset with agriculture department advisers at all.

The next section on impacts focussed more directly on the paradigm shift, and the responses of those we interviewed to the following question: *What is your opinion of the view that governments should not consider drought to be a disaster or unusual event and should not provide drought assistance to farmers?*

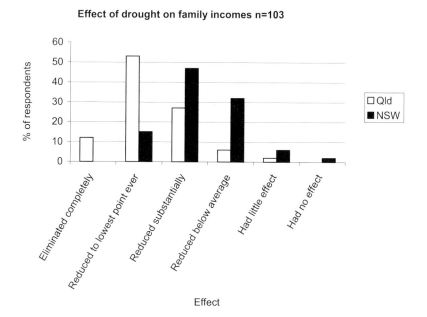

Fig. 4 Effect of drought on family incomes

The majority of those interviewed were well aware of the shift in policy. For many of them, it was an added blow in hard times. As one male sheep/wheat farmer in NSW put it: 'Drought is a disaster in a way. It's a silent disaster but something you can't stop'. A male grazier in CQ was clear about the reasons for the shift—'that was a bloody red herring to stop' [payments]. [Prime Minister] Keating kept saying 'no it is not a natural disaster. It is a 'way of life' that enabled him not to put his hand in his pocket. Purely political. Just ridiculous'.

Following analysis, five themes emerged:

- A feeling of being out of control
- Getting a fair go
- Can drought be managed for?
- When is a disaster not a disaster?
- Community impact

These themes will now be discussed in more detail by drawing on some the voices of those interviewed.

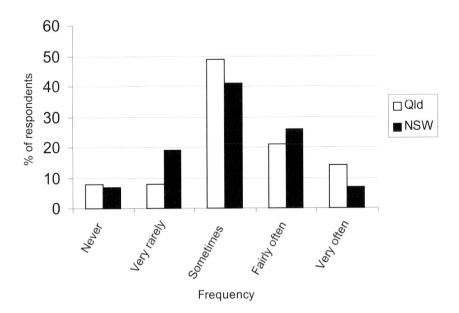

Fig. 5 Frequency of perceived personal stress

4.2 A FEELING OF BEING OUT OF CONTROL

This theme emerged as people spoke about the length of time this drought had affected them, and how no one could have planned for it lasting this long. Thus their own sense of 'disaster' can be seen as a 'loss of control' or a 'lack of autonomy' in their lives. Two key aspects of this theme act as evidence for it. First, the way in which most people compared this drought to previous events, and second, how the current event was compared with other natural disasters—most of which the government still accepted as such, and managed under Natural Disaster Relief Arrangements or Emergency Management arrangements.

One grazier had kept records on his property in Central Queensland going back 120 years. He said:

> What we have just been through is more than twice as bad as the worst drought previously recorded which was in 1902 … [In] this drought, we have had trees dying here. Those trees were here when Captain Cook

sailed up the coast [c1770] so when you have got brigalow trees dying it
is a drought—there is no doubt about it.

This drought of nearly 100 years ago was vivid in the memory of a Queensland female
grazier we interviewed. However the survival strategies of that time could not be used
today as she described.

> …my grandfather, the 1902 drought wiped him out. He had 2000 head of
> cattle at the start, enough to pay the grocery bill. He went droving for
> years, and that was enough to get him started again but [we] are not in that
> position any more.

Another female grazier from Queensland suggested that both the length of time of the
lack of rain, and the unreliability of rain when it did come, made this event a disaster.

> This drought was so widespread … a property [near here] they missed out
> on every drop. [If] we got 25 inches, they got 10 points. The poor buggers
> never got a drop of rain. Those poor things they just never had the rain.

In New South Wales, the sheep/wheat farmers agreed with their Queensland colleagues.
One female farmer said:

> …as I said things [are] out of our control. …we cannot make it rain and
> you can't make grass grow, there are so many things that are variable and
> you have no control over them.

In comparing this event to other drought events, but also to other natural disasters,
people responded by questioning why governments would choose to deem drought a
non-disaster, while floods and earthquakes were. Some comments included:

> What is the difference between a drought and a cyclone? They are both
> just as devastating. It is a disaster. It is not something that we can fully
> control yet. We can probably be like people who live in a cyclone area,
> building stronger house to protect themselves. … (Female farmer, NSW).

> I think droughts are an act of nature, of weather conditions and I think
> there are disasters in other areas of nature—a storm collapsing a building,
> or whatever [and] the government often comes to the aid of those people
> so, yes, I don't think it should be any different to anywhere else, or
> floods, or fire (Male farmer, NSW).

> [the question] is stupid. Because everybody else gets assistance if there is
> a flood or a fire. Why shouldn't the farmers get assistance? We are
> feeding the nation … (Female grazier, CQ).

> It is a disaster. The same as if you get a cyclone, that happens more
> regularly than a drought. (Male grazier, CQ).

I think [the government's position] is absolutely stupid. It is a disaster. We haven't created it. It is the same as the cyclone or earthquake. It is the elements [that are] beyond us. (Female grazier, CQ).

4.3 GETTING A FAIR GO

The next grouped theme is that of 'getting a fair go'. The 'fair go' is a highly valued aspect of the Australian tradition and identity. This egalitarian approach emerged particularly from comments of many of the women interviewed. This can be analysed as an attempt to develop a social justice ethic around the disaster of drought. Respondents again questioned why government assistance was provided for other disastrous events, but also why assistance was available when people were not in as dire straits. It was also raised as an issue given people's contribution to taxation and, through it, to the welfare 'safety net'. Many male farmers believed that the 'fair go' principle should be adhered to as agricultural enterprises kept the country balanced in terms of international trade. Finally, the 'fair go' issue was also raised in regard to the equity of distribution of any assistance. The next group of comments are stressing a desire for equity across the nation:

> ...[policy] guidelines...have to be equitable so that everyone is eligible under their own property...(Female sheep/wheat farmer, NSW).

> ...I think when they provide drought assistance they are only giving us back a bit of what we've given the country anyway (Female sheep/wheat farmer, NSW).

> I think taking away drought assistance is like taking away social security really. If you give to one you've got to give to the others (Female sheep/wheat farmer, NSW).

> [I think farmers] are more important to Australia than anyone else so I think they should help us (Male sheep/wheat farmer, NSW).

> We are producing 40% of the total income of the country, and we are the only rural country in the world that doesn't gain any assistance, or subsidy...(Male sheep/wheat farmer, NSW).

> People have to realise that a lot of money is generated on farms in the good times so in the bad times they've got to be helped and be repaid (Male sheep/wheat farmer, NSW).

> ...city people get all their job searches and allowances so why shouldn't the country people get the same? (Female sheep/wheat farmer, NSW).

> ...I think who ever thought of that idea [non-disaster] should come and live through a drought. They certainly don't know what it is all about (Female sheep/wheat farmer, NSW).

> The rural sector provides a lot of money when things are viable (Female grazier, CQ).

> ...[the government] have got their hands in your pockets when things are good. You are only getting your own money back when you put in for assistance (Male grazier, CQ).

> ...[it's] not that you want the money for yourself, for your own personal greed...(Female grazier, CQ).

There was also a call to a sense of identity and nationality in responses. The crisis had raised questions as to the place of rural Australia in the broader civic context, and the impact of the drought and the government's responses to it, challenging people's sense of themselves as citizens. The impact of the urban/rural divide discussion (see above) emerges in people's responses.

> I suppose you have to look at how important the government [sees] the whole agricultural industry ... (Male sheep/wheat farmer, NSW).

> ...there are so many other people in the community, in industry,...the needy, get government assistance. The arts. When they are throwing it around like water...and then people bellyache about farm subsidies.

> ...when everybody else is getting it, I would like [farmers] to get it (Female grazier, CQ).

> ...if there is a war here. The farmer, the boys and the farmer will have to defend their country so will the boys from town (Male grazier, CQ).

> I think basically urban Australia wants us to be their heritage and they don't want to lose us, and we really have to get in and make it (Female grazier, CQ).

4.4 CAN DROUGHT BE MANAGED FOR?

The concept of self-reliance is one with which Australian farmers are all too familiar. However the question whether droughts of this magnitude and length of time could successfully be managed raised some interesting responses.

> People can be prepared more than what they were but certainly there would still be a lot more costs and loss of production (Female grazier, CQ).

> These government schemes [are] a waste of time, we never have been able to use those, they just don't work (Male grazier, CQ).

There are three things: …interest rates…rain…commodity prices. Providing you have got two of those going for you, you can pull through. When you get the whole lot lumped together, you are in deep trouble (Male grazier, CQ).

We should manage for drought, but we have had a lot of tax deductible things taken away…(Female grazier, CQ).

Not to recognise the fact [of drought] leaves it too long. They react too slowly. They are too severe when they react and diminish the whole impact (Female sheep/wheat farmer, NSW).

4.5 WHEN IS A DISASTER NOT A DISASTER?

Definitions do matter, when whole policies are based on the meanings inherent in language. The Merriam-Webster dictionary defines 'disaster' as a great misfortune, a mishap, a calamity. Most of the respondents felt that the calamitous nature of this event rested in its longevity. Drought as a regular event *was* something that they anticipated. As one female sheep/wheat farmer put it, 'drought is not an unusual event. We live in a semi-arid area, and so it's not unusual for us to be in drought…. Six years [of it] *that* is unusual'. So the question of definition became very important, and the respondents had given this much thought.

What is considered a drought and what is not, that is the thing. When is a drought a drought? (Male grazier, CQ).

I think they need to formulate a definition of drought. A flood is a flood, a cyclone is a cyclone and an earthquake is an earthquake, but they don't act if it is over a certain level of the Richter scale. I think you need to have a definite definition…(Female grazier, CQ).

A drought is different to a dry…in situations like this and in the eighties, where it really went on and on, I think they have to recognise we are in trouble…(Female grazier, CQ).

This drought is a disaster. When you have got a drought and the poor managers are in trouble and the good managers aren't, that is not a natural disaster…but when it gets on [so that] everyone is in trouble [then it is]. (Female grazier, CQ).

[Defining it is difficult]…unless it is something very, very exceptional. Different to the norm, defining the norm is the hard part (Male grazier, CQ).

…to call what is happening in Queensland not a disaster is a joke – if that is not a national disaster I don't know what is (Male sheep/wheat farmer, NSW).

…but it is also unpredictable and this current one is excessive (Male sheep/wheat farmer, NSW).

…there are varying degrees of drought and they are affected by climatic conditions and commodity prices (Female sheep/wheat farmer, NSW).

4.6 COMMUNITY IMPACT

It is generally understood that 'disaster' implies that more than one person, or one family, is involved. We are familiar with the notion that disaster involves whole communities, sometimes whole societies. As this chapter has suggested, the paradigm shift has placed the onus on individuals and families to respond to drought—ignoring the collectivity associated with a disaster. In their attempt to define drought and resist the policy changes, our respondents talked about the impact on the whole of their communities. The inter-relationship between their own sense of calamity and that of the wider community was all too clear to them, a fact which highlighted that to them, this was what determined that it *was* a disaster …

If [producers] go off the [farm] you lose a family farm…you have got to weigh up the social costs of that…(Male grazier, CQ).

These people are feeding the bulk of the rest of the population, and if they are not there, what are they going to eat? (Female grazier, CQ).

Generally people who live this lifestyle love it whether it's good or bad, that's it and often that's all they know. If you take away their farm, you put them on the dole (Female sheep/wheat farmer, NSW).

But why should a family be pushed off, it's your home and it's your heritage. It's not like walking away from a business (Female sheep/wheat farmer, NSW).

5. Conclusions and reflections

At the time of writing, parts of Australia are still drought declared. Therefore the question of definition of what is a disaster becomes not just academic, but a matter of, literally, life or death. For many of those interviewed 7 years ago, although the past three seasons have been reasonably good, given the current drought (2002/03) they would be again worrying about the future, and whether the return to poor seasons will once again mean a struggle over the meaning of disaster.

A reading of the policies of the past two decades appear to show that those reflecting on the 1980s experience, from the early 1990s, thought that they had 'got it right' for the next event. At that time, it was argued that a whole of farm approach, supported by computer and satellite-based technologies, would show the way (White *et al* 1995, p255). However, just as the 1990s event, coming so quickly on the heels of the 1980s one, challenged this complacency, so the current (2002/03) drought challenges the policies of the 1990s. It can be said that policy development is always 'one step behind' and in practice, this makes for frustration at the farm level. Although the ideal may be a turn away from crisis management towards 'a set of policy instruments that meets the needs of farm families, their communities, the environment and the broader economy in a way that is in harmony with Australian biophysical and climate reality' (Botterill and Fisher 2003, p ix), this appears to be no closer to a reality. Waterford suggests that the 1990s policies have 'compromised' the capacity of assistance being delivered effectively and efficiently.

> Not only has this destocking affected capacities to respond quickly, but it has made administration and politics far more impersonal, far less flexible and far less responsive to individual circumstances (Waterford 2002, p4).

This compromise has intersected with a growth in social security demands, which Hancock argues has reached 'crisis proportions' (Hancock 2002, p131), and with political realignments as citizenship in rural and regional Australia reconsiders its political loyalties. How can rural families continue to cope under a self-reliance ideology? The managed risk approach, as much of neo-liberalism has tended to do, rests on traditional notions of family and community with 'predictable, linear, lifecycle related transitions'. The social policies of the future need to reassess risks as possibly being 'non-linear, episodic, multiple and recurring' (Hancock 2002, p127). For drought, a recurring reality in the Australian environment, risk management in regard to social policy needs to consider the changing demographics and increasing demands for equity by those who live outside of the 'city triangle' and whose livelihoods are connected with agriculture. As Pritchard suggests, rural Australia is a 'complex entity [and] it therefore follows that policy formulation must also be multi-dimensional' (Pritchard 2002).

Decisions regarding policy made in the 1990s at the height of that drought event are now being experienced by many who are still drought declared. They remain complex and less than adequate, and, despite some valiant attempts to enable collaboration, the demarcation between federal, state and local governments continues. Coupled with the politicised nature of 'exceptional circumstances' the transition from drought as 'disaster' to managed risk within a self-help ideology has had a major impact; in particular it has meant that 'exceptional circumstances' are now crucial in determining the severity of the event, despite the continuing concern about definitions of 'exceptional'. The decision to move from disaster to managed risk (self-reliance) also resulted in a shift from subsidies to welfare, and from public sector responsibility that originally lay in agricultural departments towards welfare and social security departments. As a paradigm shift in service support to rural families, this remains the major 'legacy' of the drought of the 1990s and the key issue to developing social policies for future droughts.

CHAPTER 6: DROUGHT, NEWS MEDIA AND POLICY DEBATE

IAN WARD

School of Political Science and International Studies, The University of Queensland, Brisbane QLD 4072, Australia

1. Introduction

On 4 December 2002, a 'Rain Train' bedecked with Channel Seven emblems departed from Sydney heading for Narrabri with the aim of distributing some 200 tonnes of donated food and Christmas toys, and thereby bringing relief to drought affected families in rural NSW. It left Sydney's Central Station in the bright glare of a media spotlight. Channel Seven's *Sunrise* morning television program, which had carried the story for several mornings, showed footage of parched rural properties, and periodically crossed 'live' to volunteers associated with the appeal, and to suffering rural families for whom the relief was intended. The Rain Train had originally been proposed by 2UE's Breakfast Program announcer Steve Price, and prompted by callers' concerns about hardship being experienced in regional areas occasioned by severe drought. Price's initial appeal to his Sydney audience to donate hampers of food for drought-affected farm families mushroomed, with the support of other Sydney media, into a more substantial relief effort supported by schools, several corporations, a bank, the NSW rail authority and a generous public.

The Rain Train appeal echoed the much more substantial, media-driven nation-wide Farm Hand Appeal which had been launched in October 2002, aggressively promoted in the pages of News Ltd papers across Australia, and given on-air support by channels Nine, Seven, Ten, WIN, Prime and ABC. Smaller in scale, the Rain Train nonetheless surely directly touched the lives of families in Narrabri, Pilliga, and other drought affected areas of northern NSW where the Country Womens' Association was enlisted to distribute donated aid. However, for all the good will involved and good achieved, the Rain Train suggests the shortcomings of news media coverage of drought. News media have a bias toward 'event based coverage' and are ill-suited to 'covering complex, complicated subjects' or slowly-developing processes (Cate 1996, pp21-22; see also Mayer 1994, p146). Unlike natural disasters such as floods and bush fires, droughts do not produce a sudden, newsworthy crisis. Thus without a specific event such as the Rain Train appeal around which to frame stories, the slow onset of drought may well go unreported.

News media are 'not simply a reflector or communication channel that plays back what it sees and hears' (Putnam 2002, p119). Indeed in every sense the Rain Train was a 'manufactured' story—a happening constructed by a news media, albeit well meaning, with the goal of generating a run of news stories about the impact of drought on rural Australia. In telling the story, reporters recycled a familiar mythology (see Stehlik *et al* 1999, p1; Wahlquist 2003, p74)—they used pictures of dusty paddocks, told heart-

L.C. Botterill and D.A. Wilhite (eds.), From Disaster Response to Risk Management, 85–98.
© 2005 *Springer. Printed in the Netherlands.*

breaking tales of battlers in the bush, of mateship and hardship, and of farmers held
hostage to an inexorable natural force. As is often the case with news stories, the Rain
Train pulled from the station and disappeared quickly from view. News stories often
have a short shelf life. The requirement that news deal with 'new' events means that
stories often attract an initial rush of attention and then disappear from front pages and
news bulletins to be replaced by fresh items.

2. Disaster stories

The familiar description of news items as stories warrants closer inspection. As
professionals, journalists have a commitment to objectivity (Schultz 1998, p133). The
ethical code of the professional association to which Australian journalists belong
requires them to 'report and interpret honestly' and to strive for 'accuracy, fairness and
disclosure of all essential facts'. News must be factual and journalists will take
professional pride in accuracy. Yet at the same time they see themselves as writing or
videoing 'stories'. To attract and hold an audience or readers' attention, news must be
entertaining as well as accurate and informative, and the best news items have a
dramatic element. Just as with good fiction, news stories will have conflict, heroes,
victims, problems and solutions, and narrate events that unfold toward a conclusion. As
Auf der Heide (1989, p216) observes, natural disasters tend to 'offer all these
characteristics, and for television they present the additional advantage of [providing]
great attention-grabbing visuals.'

Disaster mitigation activity is rarely newsworthy. But the actual occurrence of a natural
disaster naturally makes news. Seen from a 'journalistic point of view' natural disasters
'have all the ingredients for the perfect media event' (see Aufe der Heide 1989, p216;
Bolduc 1987, p12; Mayer 1994, p142). Earthquakes, storms, earthquakes, floods and
fires unexpectedly disrupt everyday life, and often attract news coverage. This is most
sharply demonstrated by the so-called 'CNN effect' and the propensity for television
news to cover otherwise distant countries only when disasters such as famine or
earthquakes provide 'dramatic news coverage of suffering people' (also see Natsios
1996; Robinson 2002, p175). Given their audience reach and role in representing and
reconstituting events, it is not surprising that the manner in which news media cover
disasters has been the subject of some scholarly examination, albeit limited (see
Quarantelli 1989, p8). In the immediate locality of a disaster the news media are often a
chief means of distributing information about the extent of damage and appropriate
safety measures to be taken. More broadly, the news media, especially television, are
'the most important source from which the public obtains information on disasters'
(Aufe der Heide 1989, p217). But news reports do not faithfully mirror the real world.

Journalists may place great store upon accuracy and factual reporting. But news reports
are necessarily brief and selective, and 'when selection is necessary, distortions are
inevitable' (Graber 2001, p121). Despite the best attempts of journalists to accurately
report 'the facts', this is often difficult to do, especially in the case of disasters that
rapidly unfold. Sood, Stockdale and Rogers (1987, p28) observe that, for journalists, 'a
natural disaster is usually also an information disaster'. Often much initial reporting is

done on the run. In the immediate wake of a disaster incident, journalists can have great difficulty in establishing its cause and meaning, and the extent and scope of damage (see Mayer 1994). To begin with, they will most often lack expertise and have much 'difficulty evaluating the technical aspects of disasters' and identifying reliable sources (Aufe der Heide 1989, p233). Invariably there will be different and competing eyewitness accounts of what happened, of the number of casualties, and of the extent of property damage. Normal communications infrastructure may be damaged. The congregation of journalists at the scene of a disaster incident can itself be a cause of disruption and add to confusion (Ewart 2002, p4; Sood *et al* 1987, p32). The police, health workers, fire-fighters and other official sources on whom journalists might otherwise rely for 'hard' information will be preoccupied with dealing with the disaster's aftermath and be unavailable, or themselves unaware of the full extent of damage. In this context there is every prospect that the news coverage may not be accurate and that the news media will inadvertently 'misinform, distort and misfocus attention' (Cate 1996, p19; see also Elliott 1989, p167). This is a particular risk where reporters face 'fierce competition' and the 'associated pressures' to get the story out (Ewart 2002, p3), and where 'distance and time constraints combine to reduce the opportunity for first-hand evaluation and thorough fact-checking' (Cate 1996, p19).

It is a commonplace complaint—though disputed by some (see Goltz 1984; or Quarantelli 1989, p14)—that disaster news too often falls back upon enduring myths (Wenger and Friedman 1986; Wilkins 1986). These myths include the belief that when disaster strikes people will panic, refuse to move to safety, or engage in looting, or that disasters escalate the risk of the outbreak of contagious disease. Myths associated with natural disasters such as floods and storms include the idea that nature is powerful and prone to randomly wreak havoc and human suffering. Television in particular is likely to focus upon images of chaos and disruption and the plight of victims. As a consequence, news coverage of natural disasters can convey a sense of hopelessness, depicting people as powerless in the face of, and ultimately hostage to, the random force of nature. News stories of this kind, especially when they contain graphic images, can create a considerable public sympathy for nature's victims.

What is clear from the accumulated, careful study of how journalists gather, report and produce news is that news work is a structured, rule-bound, routinised activity (also Tuchman 1997; see Ward 1995, pp102-11). The employing of myths in disaster reporting can be seen in this context. Drawing on Meadows' (2001) account of the cultural practices embedded in journalism, Ewart (2002, p2) describes how the daily routines of journalists entrench 'accepted practices' including particular narrative techniques or ways of interpreting information and making sense of stories. Journalists, in covering a story, will 'select approaches from their pre-established repertoire' and adhere to what Ericson, Baranek and Chan (Ericson *et al* 1987, p348) describe as a 'vocabulary of precedents ... [and] previous exemplars [which] tell them [what] should be done in the present instance'. In this vein Jemphrey and Berrington (2000, p469) suggest that 'much journalistic practice is routine, with stories … reported in accordance with pre-constructed news templates'. They argue that disaster reporting is no different. Disasters may disrupt normal news routines, but as Vincent, Crow and Davis (1997, pp355-6) show, journalists cope by applying 'particular narrative devices and story construction strategies'. For all the disruption and difficulties that reporters face because natural disasters are also information disasters, in the immediate aftermath of a disaster event, journalists will routinely seek 'to establish the number of casualties, the apparent cause and the identity of those to whom blame can be apportioned' and then supplement this with information about victims and survivors and expert comment 'from the police and other emergency services' (Jemphrey *et al* 2000, p468). Put another way, journalists appear to consistently 'frame' reports of disasters in much the same way.

3. Drought as disaster news

With the 1989 removal of drought from the umbrella of Commonwealth-state Natural Disaster Relief Arrangements and the subsequent introduction of the 1992 National Drought Policy, policy makers attempted to jettison a view of drought as a 'natural disaster' requiring governments to provide relief to its victims. However, for the news media drought is ultimately just another natural disaster story to bring to their audience. Even though drought may unfold very differently from 'rapid onset' disasters and, without a 'sudden crisis' to draw media interest (Mayer 1994, p146), not occasion the same at-the-scene 'initial rush to gather and disseminate news' (Ewart 2002, p4), news coverage of drought retains many elements of disaster reporting. As with natural disasters such as floods and fires, in the case of drought we might also expect to see that local and national media—each with a different audience and the former living within communities effected—will also provide a different news coverage (see Aufe der Heide 1989, pp232-3; Jemphrey *et al* 2000, p471). As with disaster reporting generally, news coverage of drought is prone to exaggeration where 'the facts' cannot be reliably established. Inevitably there must be an uncertainty about how long drought will persist, and about how severe its economic and social consequences will ultimately be. Such uncertainty allows exaggerated estimates of the severity and impact of drought. As with other natural disasters, news coverage of drought will also have its share of 'human interest' stories focussing on the plight in which its victims find themselves and their courage in the face of adversity.

Yet there are marked differences between droughts and the bushfires, storms and floods that periodically disrupt life in rural and regional Australia. Droughts slowly emerge, and then linger. Furthermore, unlike other 'rapid-onset' disasters, droughts tend neither to have 'rather clear cut boundaries', nor to occur at 'definite points in time' (Sood *et al* 1987, p29). Droughts do not happen in a confined or specific geographic location. Nor do occurrences of drought have a defining, cataclysmic event to provide a 'peg' around which news coverage can be constructed. Drought is a regular and persistent visitor to Australia's shores. Its common occurrence may mean that drought only becomes newsworthy when it reaches particularly severe levels. Although droughts clearly have features that distinguish them from other rapid-onset natural disasters, there are some lessons from the wider study of news reporting of natural disasters that still hold.

In reporting droughts, journalists will not face the disruption of routine, frantic competition for the story, widespread confusion about what has happened, or the unavailability of police and emergency services sources, all of which are commonplace obstacles to reporting in the immediate wake of fires, floods or similar natural disasters. Nevertheless a drought can also be an information disaster for journalists assigned to cover it. The causes of drought are complex, and difficult to grasp and to put into words within the confines of a brief news item. Its effects upon economic and social life are gradual, difficult to measure, and most often borne by communities distant from metropolitan newsrooms. Journalists faced with a complex, ill-defined phenomenon such as drought are likely to turn to 'pre-constructed templates'. Droughts may be rather different from other forms of natural disaster, but the manner in which drought will be reported ultimately owes much more to the routinised ways journalists report the news than to the nature of drought.

It is now well understood that news work requires journalists to deal with unpredictable events in routine ways. By definition, news deals with novel and therefore unanticipated occurrences, and this obliges news organisations to develop 'some routine method of coping with unexpected events' (Tuchman 1997, p174). This routine will allow newsrooms to deal even with large scale, rapid-onset disasters (see Berkowitz 1997, p373; and Vincent *et al* 1997). Simply, journalists learn 'convenient ways of pigeonholing information' or of 'framing' events that allow them to make sense of, and to recount, rapidly unfolding developments (Ward 1995, p112). News frames are 'little tacit theories about what exists, what happens and what matters'. They are 'clamped over' or made to fit sets of events with numerous potentially 'mentionable' details (Gitlin 1980, pp6-7). Resultant news stories will play down or overlook details that fall outside the selected news frame. Journalists who report drought will approach the task in this same way. However if using news frames is 'an unavoidable feature of news work', journalists covering drought will still have a 'considerable discretion in selecting which particular frame to apply' (Ward 1995, p112). How they choose to frame drought stories—the tacit little theories that they develop and the details they call attention to—may well have significant consequences. News frames, as Putnam (2002, p120) notes, 'highlight or promote particular definitions and interpretations of situations'.

Studies of disaster reporting suggest that the manner in which journalists on the scene frame news stories can influence how emergency workers and the public alike understand what has happened. Sood, Stockdale and Rogers (1987, p39) emphasise that news coverage of disaster incidents involves more than 'simply the coverage of the "facts"'. Reporters draw attention to 'certain elements' of a disaster event and these elements then become 'part of the rhetoric of future public and policy debates'. Indeed Quarantelli (1989, p14) observes that a 'strong theme' in many studies of news coverage of disasters emphasises that news does 'not reflect reality' but instead defines how a disaster, its causes and consequences are understood. How the media frame their coverage of drought may well have public policy implications (see Wahlquist 2003, p85). For example, in their analysis of the drought coverage provided by a regional Queensland newspaper, Mules, Schirato and Wigman (1995, p246) point out that 'farmers and their representatives' sought to 'present the drought as a natural disaster' in order to legitimate their claims for financial assistance from government. More pointedly, Gow (1994-95, p7) argues that the manner in which news media report drought can 'build pressure upon the government to be seen to be "doing something"' which will provide relief for drought-stricken rural producers, and that this has, in the past, produced knee-jerk policy responses which are not in the long-term national interest.

4. The media's role in framing policy debate

Gow's (1994-95, p11) particular complaint that sensationalist news coverage of drought in Australia has been a 'hindrance to an informed debate about what can be done to minimize the effects of drought' warrants closer inspection. It suggests that the manner in which the news media frame their coverage of drought can directly influence the ways in which policy makers understand and respond to the issue. It is a commonplace suggestion in the literature on natural disasters that the manner in which news media report disasters can have an impact on public policy (for example Birkland 1996; Cate 1996; Shattuck 1996), although it would seem an overstatement to argue that the power of news media is such that 'few causes or events, no matter how dramatic they are or how many people are involved, motivate powerful governmental or institutional responses until captured by the cameras of the press' (Cate 1996, p18). The relationship between news media coverage of an issue and the response of policy makers is likely to more complex than this suggests.

In a democracy such as Australia, legislators will understandably keep a weather eye on public opinion with the next election in mind. Hence it is likely that the manner in which an issue is reported will influence the priority policy makers will attach to it. Political communication scholars have long recognised that news media can have an agenda-setting role. There is a substantive body of evidence comprising more than 350 studies which point to the capacity of news media to draw public attention to particular issues by giving them prominent and frequent attention (McCombs and Reyolds 2002, p3). Rogers and Dearing's (1988, p557) much-reproduced model of the agenda-setting process does suggest that any linkage between media coverage of an issue and the priority policy makers attach to it will necessarily be complex. Policy makers do not simply, nor only, respond to shifts in public opinion. In some cases they may react directly to news coverage of an issue, in effect treating the emphasis given it in the news as a measure of its importance. But equally, policy makers may be driven by lobbying or by other communications between elites occurring outside of the media, or by economic trends, international pressure, or other 'real-world' indicators of the importance of an issue. However, this complexity notwithstanding, news media coverage of an issue 'seems to have direct, sometimes strong, influence upon the policy agenda of elite decision makers' and in some cases even upon policy implementation (Rogers and Dearing 1988, p580).

It is true that 'we know less about how public policy debates are shaped than we might like to admit' (Terkildsen et al 1998, p45). The nexus between the news and policy making is not as well understood as it needs to be. To begin with, the news audience is fragmented. It is divided between print and broadcast media outlets with different news priorities and ways of handling news stories. This latter point is often made in studies of disaster reporting (Quarantelli 1989, p14). Television news demands compelling images to the point where the availability of striking video may determine whether or not a disaster incident is reported, whereas print news outlets are less prone to this 'CNN effect' and better suited to conveying complexity. Just as there are differences between news media, the nature of news stories themselves can differ. Some will attract an initial rush of attention and then quickly disappear from front pages and news bulletins to be replaced by fresh, new stories. Others may unfold slowly, generate ongoing coverage, and thus potentially have a rather different impact on public opinion and the attention of policy makers.

Even where news coverage stirs widespread public interest in an issue (such as drought) it is not axiomatic that policy makers will be moved to act. This may be explained by Downs' (1972) observation that issues pass through attention cycles. His model describes a 'pre-problem stage' during which there is little public (or media) interest in an issue. This is followed by a second, 'alarmed discovery' stage, which is often triggered by extensive news coverage of a disaster incident and characterised by widespread expressions of public concern and accompanying demands for governmental action. Thereafter issues tend to pass into a third stage during which the public debate centres on the costs associated with proposed policy solutions, and then a further phase in which public interest (along with attendant media coverage) wanes. The final stage in Downs' issue attention cycle is one in which the public and the media lose interest in the issue, and in which new problems come to dominate public debate. It is a mistake to imagine that policy making proceeds in an orderly, linear fashion from the identification of a problem to the development and implementation of an appropriate solution. As has been widely recognised, policy making is far more politically messy and far less ordered than this allows (Parsons 1995, p433). However, Downs' model does allow us to see that the effect of news media coverage of an issue may vary over time, and that the media may be far more influential at some stages of the policy-making process than others.

A clear implication of Downs' issue attention cycle is that, at particular stages of the policy-making process, news media can play a significant part in drawing an issue to the attention of policy makers and in driving them to attend to it. More recently it has been suggested that the particular manner in which news media frame the coverage of a policy issue may actually shape the policy solution that is ultimately adopted. A considerable power over policy making resides in 'the process whereby problems are constructed and articulated' (Parsons 1995, p180). Rein and Schön (1993, pp146-7) point out that framing an 'amorphous, ill-defined, problematic situation' is a way of making sense and acting upon it. Framing a problem in one way, and not another, leads to making 'different interpretations of the way things are' and hence supports 'different courses of action concerning what is to be done, by whom, and how'. As Parsons (1995, p85) argues, public policy problems have 'to be defined, structured, located within certain boundaries and given name', and how this happens often determines the policy solution that ultimately emerges. Thus drought defined as a natural disaster and understood as an inexorable, uncontrollable natural force driving rural producers into severe economic difficulty[1] constructs a policy solution that involves channelling relief payments to suffering rural producers. But drought understood as a naturally reoccurring phenomenon points to a very different policy solution of assisting farmers to adopt 'drought proof' or sustainable farming and land management practices.

Edelman (1988, p104), recognising that naming and framing a policy problem is likely to determine what policy solution is likely to be put in place, argued that the 'critical element in political manoeuvre for advantage is the creation of meaning'. Indeed a key political tactic must always be 'the evocation of interpretations that legitimate favoured courses of action'. Obviously news media are one important arena in which competing political interests can manoeuvre for advantage in a contest to determine who gets what from government. Equally, the ways in which editors and journalists understand, describe and report a problem or issue can assist or encumber the efforts of an interest group to influence policy. It follows that, in order to understand the making of drought policy in Australia, we need to look more closely at how drought is reported, especially in the national news media which have the largest audience reach and whose urban viewers and readers will be unlikely to have first-hand experience of drought on which to draw.

5. How the national print news media cover drought

Even though Australia is regularly visited by drought, there have been few studies of the coverage of drought by Australian news media (for example, Mules *et al* 1995; Wahlquist 2003). None has looked systematically at the manner in which the national press covers drought. However, the Lexis database now contains the full text of back issues of major Australian newspapers spanning back to the latter 1990s. It is now a relatively straightforward task to establish the frequency with which news items and commentaries referring to drought appear (although 'drought' has such a surprising currency as a metaphor for lean times in the sporting arena and elsewhere that a careful inspection of each story retrieved using this key word is required in order to eliminate entirely unrelated news items). In Figure 1 the numbers of drought stories appearing each month in the *Sydney Morning Herald* (from January 1997) and *The Australian* (from January 1996) are separately plotted until the end of 2002. These are two major 'journals of record'. The pattern of coverage that they display follows a remarkably similar trend which is likely to be at least broadly representative of the wider metropolitan newspaper coverage of drought.

---

[1] In this context it is worth noting that the very notion of a natural disaster deflects attention from associated policy issues. In fact, as Benthall (1993, p13) suggests, so-called natural disasters 'nearly always include a human element'. Thus the damage wrought by earthquakes can be a by-product of inappropriate building practices, and that caused by floods a result of poor land management policies allowing settlement on vulnerable flood plains. Similarly, bush fires may result from the failure of park authorities to reduce forest-floor fuel with preventative, controlled burning programs.

Figure 1 suggests that the pattern of news coverage associated with drought, while seemingly cyclical, will be very different from that given to floods, bushfires and rapid-onset disasters that will generate an initial rush of stories but then fade from the news. Sustained or severe periods of drought can be expected to trigger more news reporting. But it is not unusual for some part of the Australian continent to face drought conditions. This may be one reason why drought never slips entirely from view, as the peaks of news interest in drought are set against a continuing background coverage of the problem. Not unexpectedly during 2002, news coverage soared as Australia endured a particularly widespread and prolonged period of drought.

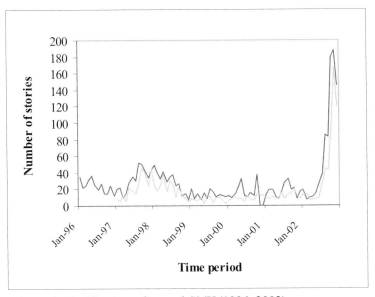

Fig.1 Drought stories in *The Australian* and *SMH* (1996–2002)

The frequency with which newspapers mention drought begs the more important question of how news media understand, and tell the story of, drought. If the frequency of news coverage of an issue draws it to the attention of the public and thus policy makers, the particular manner in which news media frame that issue may influence how policy makers understand and respond to the problem. As we have seen, writing a decade ago and very much in this vein, Gow (1994-95, p7) argued that the particular way news media understood and reported drought was an obstacle to making good public policy decisions. Clearly we need to examine not simply the frequency with which news media report drought but the way in which episodes of drought are framed.

Table 1 sets out the results of a content analysis of a randomly selected sample of 100 news stories, all collected via the Lexis database from the stories mentioning drought that appeared in *The Australian* newspaper during the second half of 2002. Admittedly these data may not fully mirror news coverage of drought. For example, it might be expected that, with a prolonged drought well established in the latter part of 2002 across much of Australia, there would have been fewer stories about the declaration of drought-affected areas and more about drought relief measures (such as the Farmhand and Rain Train appeals). Nonetheless the data displayed do provide some insight into how Australia's chief national daily newspaper reports drought. It is noteworthy that almost half of the stories (48%) that mention drought conditions will have quite another focus. Most often these are stories about the state of the wider economy that draw attention to the economic cost of drought. Indeed this finding may well help explain the continuing background of news stories mentioning drought shown in Figure 1, given that a great deal of news space is routinely devoted to the economy. At first glance this focus on the economic impact of drought might also appear to confirm Gow's (1994-95, p7) complaint that a constant theme within news reporting is the 'assertion that drought costs the nation billions of dollars'—a way of framing drought which he believes must predispose policy makers to provide drought relief subsidies to farmers.

Table 1. Analysis of the content of drought stories in *The Australian* (June-December 2002)

Story attributes	%
Incidental mention of drought as part of another story	48
Parched paddocks or similar physical manifestations of drought	15
Unforgiving climate and/or El Niño and prolonged dry weather	25
Struggling farmers facing hard times	16
Reporting the declaration of drought-affected areas	14
Provision of or fundraising for drought relief	18
Voice given to a farmer lobby group	10
Need to drought-proof or change farming practices	13
Impact of drought on city water supplies	8
Impact of drought on the national economy	43
Overseas drought stories	9
	n=100

However, Table 1 does show that a further feature of newspaper coverage of drought appears to be its diversity. News coverage is clearly not driven simply by rural or farm lobbies whose voices are heard in but a relatively small number (10%) of stories. Nor is coverage of drought always focussed on rural Australia—8% of stories sampled raised the threat that drought posed to urban water supplies. Just 16% describe struggling farmers facing hard times. Only a relatively small proportion (18%) focus on the need for or provision of drought relief, and a similar number (13%) take up the issue of rural producers needing to adapt their water use and farming practices. Here there appears to be little empirical support for Gow's (1994-95, p7) view that the print media 'have a field day trying to outcompete each other' with 'sensationalist' reports of the negative impact of drought (which, he argues, induces public sympathy for struggling farmers and skews policy debate towards the provision of drought relief.)

There is evidence that news coverage of drought does have a mythical element (as might be expected given what is known of disaster reporting generally). For example, the idea that rural producers are exposed to the vagaries of an unforgiving or unpredictable climate occurs (in one or another form) in twenty-five percent of stories examined. Fifteen percent of stories sampled refer to parched paddocks, failed crops, cloudless skies or other physical manifestations of drought. Clearly these mythical elements are not the central motif of the majority of news stories mentioning drought. But the fact that many stories mentioning drought are not stories written specifically about drought does point to the more commonplace use of disaster myths about the power drought. A close examination of individual stories that do specifically focus on the power of nature to wreak havoc and the heart-breaking struggles of farmers plunged into hard times by the failure to rain. For example, on December 2, 2002, *The Australian* carried a page 2 story headlined 'drought aid risks handout mentality'. It recounted the concerns of rural commentators that the federal government's newly announced extension of drought relief might 'discourage farmers from preparing for droughts'. Yet the accompanying photograph of a WA farmer standing amidst a ruined farmhouse and surveying his failed crop invokes the myth of farmers battling difficult drought conditions and carries a different connotation. This is the picture of a battler deserving and in need of assistance.

On December 9, *The Australian*, to illustrate a story of a small north-western NSW rural community entirely dependent upon water carted in, carried a photograph of a lone figure standing under a cloudless sky on the cracked, dry floor of the Tiboora Dam. On December 17 the same paper reported that the Victorian government had stepped up relief payments to drought-affected farmers amidst complaints that 'the federal Government's drought assistance package is too restrictive'. It illustrated this story with a photograph of the Premier and a beleaguered farmer framed against a clear sky and forlornly inspecting a 'ravaged barley crop'. On December 23, *The Australian* carried a photograph of Queensland farmer, her son and grandchildren on the back of a utility surrounded by 'painfully lean' sheep in a bare paddock. The accompanying story told of a family 'doing it tough' at Christmas. Each of these stories also draws upon the mythology of drought—upon the idea that farmers and their families are ultimately hostages to drought and victims of natural forces beyond their control. Lexis is text-based and it is not possible to calculate how many stories about drought also carry similar pictures.[2] But drought stories containing images of this kind do suggest how, when required to visually portray something as complex and ill-defined as drought, news media may fall back onto some familiar templates grounded in myths associated with drought.

Gow's (1994-95, pp7-8) particular complaint is that 'emotionally moving' *images* used in reports of drought trigger 'sympathy for the plight of farmers', and he points to the role of television in building pressure on governments to 'do' something. Unfortunately television news cannot be retrieved as easily and conveniently as it is now possible to recover newspaper reports for content analysis and, thus far, there have been no studies of how Australian television covers drought. However, it is unlikely that TV news will

---

[2] A physical inspection of microfilmed back copies of *The Australian* identified five such photographs for the month of December 2002, and a further six that appeared with news stories in the prior month of November 2002.

reflect either the volume or diversity of coverage found in *The Australian*'s pages. Given the difficulties in capturing a complex phenomenon such as drought in pictures, it is highly likely that television news will also reproduce the images of cloudless skies, dusty paddocks, starving animals, failing crops and dry dams and battling farmers, all of which are suggested by the particular disaster myths associated with drought. Print media coverage of drought may be more diverse and generally less emotive than Gow allows. But it is possible that television coverage of drought as natural disaster may still prejudice public understanding of drought in the manner he suggests. The scale and generosity of public support for the Rain Train and Farmhand appeals (Wahlquist 2003, pp76-78) points in this very direction.

6. Exceptional circumstances?

During the 1990s the federal government sought to shift drought policy away from the provision of crisis relief toward rural adjustment and the longer-term encouragement of drought resistant farm management practices. In effect it sought to redefine drought, not as a natural disaster requiring relief payments to hard-hit farmers, but as a normal 'business risk' which rural producers must anticipate and factor into their business plans. At least in part, the 1992 National Drought Policy was intended to insulate policy makers from the political pressures that inevitably accompanied drought to make 'band aid' payments to badly affected rural producers. Yet the news media appear to have only partly accommodated this switch in policy. Much—though by no means all— news coverage of drought still frames it as natural disaster, and this is accentuated by media-backed appeals such as Farmhand and the Rain Train aimed at raising and distributing relief to suffering farm families. Nevertheless it is not clear that news coverage of drought does, as its critics suggest, build public sympathy for drought-affected farmers and thus generate a pressure upon policy makers to provide them with subsidies that actually reward unsustainable farming practices.

The lesson of Birkland's (1996) study of the impact of hurricanes and earthquakes on the US Congressional agenda is that, while both forms of disaster attract intensive media interest, policy pertaining to earthquakes and hurricanes is made in each case in a different policy environment. He shows that in each case, different legislative committees, interest groups and professional communities are involved policy decisions. The wider point would seem to apply to the making of drought policy in Australia. Although the way in which the media frame drought news has the potential to shape public opinion and thus generate demand for a particular policy solution, ultimately the farm lobbies and other interests who succeed in establishing themselves as stakeholders will shape drought policy. Present government policy allows for an Exceptional Circumstances Relief Payment to be made to farm families in exceptionally hard-hit drought-stricken areas. As the National Farmers Federation (National Farmers Federation 2002, pp7-8) has recently observed, lack of 'a clear definition of what constitutes an exceptional circumstance' inevitably produces conflict between 'those who are seeking support and those who decide if it is to be granted', and thus disputes 'between State and Commonwealth Governments', farmers and their representatives. In some cases—say the publicity generated by the Farmhand and Rain Train appeals—it is possible that the news media coverage of drought might sufficiently stir public sympathy to tip the balance within this policy environment in favour of those seeking 'exceptional circumstances' support. But in the normal course of events it is difficult to see the general news coverage of drought having any particular influence.

However, in the longer term the news media may have a subtler effect. Quarantelli (1989, p14) points out that a recurrent theme in the literature is that media reports of disasters 'considerably determines what comes to be or not to be defined as a … disaster'. In a recent policy paper the NFF (2002, p8) notes that 'in policy terms, accepting drought as a disaster suggests' the need to deal with drought under the National Disaster Relief Arrangements agreement between the Commonwealth and states. On the other hand, 'if drought is not a disaster, the policy issues relate to the type of specific drought response programs that are needed'. Simply put, the framing of drought as a form of natural disaster has broad or long-term implications for drought policy. If this is true, and policy makers do respond to the way an issue is covered by news media, then the media's evident reluctance to abandon framing drought as natural disaster may continue to subtly check the efforts of policy professionals who would redefine drought as a risk individual farmers should plan for and carry.

CHAPTER 7: AT THE INTERSECTION OF SCIENCE AND POLITICS: DEFINING EXCEPTIONAL DROUGHT

DAVID H. WHITE
ASIT Consulting, 20 Fauna Avenue, Long Beach, NSW 2536, Australia

LINDA COURTENAY BOTTERILL
National Europe Centre, 1 Liversidge Street (#67C), Australian National University, ACT 0200, Australia

BRUCE O'MEAGHER
Consultant, 169 Wattle Street, O'Connor, ACT 2602, Australia

1. Introduction

Australia introduced the National Drought Policy (NDP) in 1992 based on principles of self-reliance and risk management (see Chapter 4). The NDP was an agreement between the national and state governments which sought, for the first time, to deliver a nationally consistent approach to drought policy. Based on these principles, the governments agreed to provide additional support for drought-related research and development and improved farm management practices and to reorient their direct farm assistance policies away from ad hoc assistance to "farmers in trouble", i.e., towards farm business and welfare support directed at increased self-reliance and risk management based on objective, science-driven decision making.

Against this background, the NDP included recognition that occasionally farmers face rare and severe events that are outside the preparedness strategies of even the best managers. These events were classified as "exceptional circumstances" (EC). The declaration of EC provided a trigger initially for the provision of farm business support and after 1994 for farm family welfare assistance to individual farm units by the national government. This assistance comprised extremely generous access to components of the welfare system (including income support) and generous interest rate subsidies previously denied. Although the state governments had agreed to phase out the provision of drought-related support, not all states have done so by the time of writing.

From the outset, determining the precise spatial dimensions of an exceptional circumstance was a significant challenge, in part because of the limited relevant data and limited resources and the tight timeframes imposed for drought assessment. Nevertheless, although not without its critics, there was increasing agreement towards the end of the 1994-95 drought that the processes adopted and the decisions made represented workable outcomes based on reasonably robust decision making. Over the years since then (and especially following the onset of widespread drought conditions in

99

L.C. Botterill and D.A. Wilhite (eds.), From Disaster Response to Risk Management, 99–111.
© 2005 *Springer. Printed in the Netherlands.*

2001), the EC declaration process and welfare element of the NDP has become increasingly politicized, in spite of ongoing developments in scientific methods and approaches which, theoretically, should have led to improved objectivity in the declaration of exceptional circumstances drought.

This chapter explores some of the factors which have had an impact on the attempt to establish an objective, science-driven approach to welfare support under EC. It describes recent scientific advances and how they could have contributed and might yet contribute to developing policy measures which have the potential to improve the objectivity of any declaration process. In particular, we discuss how the highly political nature of drought policy has prevented a more objective approach from being implemented and suggest some alternative approaches which, while recognizing the fundamental constraints of the political system, can benefit from the use of science in determining drought policy responses consistent with the objectives of the NDP.

2. The importance of welfare support

When the National Drought Policy was first announced it focused entirely on farming as a business undertaking. The only welfare component of the original NDP package was aimed at encouraging non-viable farmers to leave the industry. In 1994, with the introduction of the Drought Relief Payment and other significant welfare benefits, and more particularly after the change of government in 1996, the focus of the NDP increasingly focused on the human (especially the perceived welfare) impacts of drought. Indeed, in 1999 the Commonwealth and State Ministers announced the phase out of drought-specific business support (not that this has actually happened) and that "the purpose of EC assistance is moving away from business support (through the phasing down of interest rate subsidies) with a greater resultant emphasis on EC as a welfare measure (family income support)" (ARMCANZ 1999a, p60). In this context, the broad principles of self-reliance and effective risk management underlying the NDP were seriously compromised.

There is plenty of room for debate about whether there should be special access to the welfare system for the farm sector. Indeed, the authors themselves disagree on this matter. The question of the appropriate delivery of farm welfare support is beyond the scope of this paper; however, we note that the inclusion of welfare measures in the existing policy framework complicates drought policy. The authors agree that the delivery of welfare support is, and should be, a separate concern from drought relief. But, that debate is not pursued in this paper. Instead, we turn our attention to how the NDP principles have been undermined and to approaches to giving them a more important role within the drought policy debate in the future.

3. Science and drought declarations in Australia

Drought may be simply defined as below-average rainfall that restricts typical plant growth for agricultural production. However, it is difficult to identify and measure drought by the quantity of rain alone, as the timing and frequency of precipitation as well as soil type, topography, and land management practices will affect plant responsiveness and the effectiveness of the rainfall received.

At an aggregated level, a reasonably high correlation has been found between published reports of the incidence of drought based on production criteria and with annual rainfall in the first decile (for example Gibbs and Maher 1957; Smith *et al* 1992). However, there can still be many circumstances where the rankings of droughts based on rainfall data alone differ from droughts based on pasture or crop yield (for example Keating and Meinke 1998; Stafford Smith and McKeon 1998; White *et al* 1998). Agricultural droughts must therefore be distinguished from meteorological and hydrological droughts (Wilhite 1993). Numerous other indices have been proposed based on rainfall data or soil moisture models, though most have been found too limiting for the purposes of ranking agricultural droughts as the basis of government policy intervention.

Six criteria were proposed in 1994 for determining whether an area qualified for exceptional circumstances assistance (White *et al* 1998). These were meteorological, agronomic, environmental, water supply, scale of the event and net farm income. Emphasis was placed on rainfall and effectiveness of rainfall, it being acknowledged that agricultural droughts did not necessarily coincide in occurrence or severity with meteorological droughts, even though they clearly are related. Although politicians and politics were not removed from the drought policy arena, a key objective of these measures was nevertheless to depoliticise drought policy by providing for objective, science-driven decision making.

Objective assessments of the duration, extent and severity of droughts must take into account the prevailing climate and agricultural systems in different parts of the country. Thus a 15-month drought in the high rainfall zone may have a comparable impact to a five-year drought in the semi-arid zone. The timing of rainfall events and associated soil temperatures can be crucial in determining, for example, whether annual pastures in the temperate zone germinate and their seedlings survive and actively grow following autumn rains. This can be crucial in determining when specific droughts commence and cease. These complications were factored into the DEC process, although there was debate initiated by interest groups concerning how relevant they were.

The EC decision making process is characterized by short and changing time frames, a broad variety of agricultural systems, multi-disciplinary information, and analyses at a regional scale (Walcott and Clark 2001). Reliable drought assessments are therefore highly reliant on access to climate data, agronomic field data, remotely sensed data from satellites, farm surveys, and output from agronomic models. Different tools are better suited to different areas and agricultural systems. Although skilled analysts are able to integrate a wide diversity of information using available tools to make credible judgments so as to rank the severity of droughts over say the past 100 years within a region, the complexity of the task has made the process less transparent than the community demands.

Tools for assessing and ranking droughts have developed dramatically in recent years. Rainfall records may be analysed for specific location or over different spatial scales using a range of decision support scales. Software for analysing data from specific weather stations includes an Excel spreadsheet (Bedo 1997), MetAccess for manipulating and analysing daily data (Donnelly *et al* 1997) and Australian Rainman (Clewett *et al* 2003), which focuses primarily on monthly rainfall records. A more recent development is Rainman Stream Flow (Chiew *et al* 1998; Clewett *et al* 2003) which assesses how rainfall events lead to variability in stream flow.

Digital maps that present rainfall received over specified periods ranging from days to years have been produced by the Queensland Government (http://www.longpaddock.qld.gov.au) and the Bureau of Meteorology (http://www.bom.gov.au). The maps display this information as either absolute precipitation amounts, or as deciles relative to the long-term records, and are accessible through the Internet. The Rainfall Reliability Wizard developed by the Bureau of Rural Sciences (BRS) (Laughlin *et al* 2003) enables similar maps to be displayed on a personal computer, including assessing the likelihood of rain over particular growth periods. Access to historical climate data for any point within Australia has been facilitated through the development of a data drill that contains numerous interpolated climate surfaces (http://www.bom.gov.au/silo) (Jeffrey *et al* 2001).

Assessment of agricultural droughts requires that the effectiveness of rainfall and the significance of temperature type be determined. This has been done by overlaying maps of mean and minimum May temperatures across New South Wales over corresponding information of rainfall received so as to determine whether meaningful germination and growth is likely to have occurred following the onset of autumn rains (White 1998). A more generic and favoured approach is to use models of grassland (Donnelly *et al* 1997; Stafford Smith and McKeon 1998; White *et al* 1998) and cropping systems (Keating and Meinke 1998; Stephens 1998). Such models overcome the need for predefining growth periods and moisture-temperature interactions in that these are implicit in their operations.

Aussie GRASS (Carter *et al* 2000) incorporates an extensively tested grassland model within a Geographic Information System. Developed initially as a prototype of a national drought information system, it has been extensively tested throughout Queensland, and to a greater or lesser extent in most other States and the Northern Territory (Hall *et al* 2001). It has been of considerable assistance to national drought assessment since 1996.

BRS is developing a spatial system to handle climate information and estimates of pasture production based on the relatively simple GROWEST model (Fitzpatrick and Nix 1970; Brinkley *et al*, in press). It has yet to be established whether this model, based on a single soil layer, weekly time steps, and indices of plant growth relative to soil moisture, temperature and solar radiation, can perform as credibly as the more complex models described above. This will almost certainly require extensive testing at the State and Territory levels.

Assessments of the agronomic and environmental impacts of drought have also made use of industry data, including wheat forecast and yield data from the Australian Wheat Board, and yield data from the Australian Bureau of Statistics. The latter could be compared with long-term trends, as illustrated by Hamblin and Kyneur (1993). Maps provided from farm surveys in some States gave an indication of changes in livestock numbers, the availability of pasture cover and supplementary feed, and stock water supplies. Maps were also provided that highlighted the vulnerability of soils to wind and water erosion.

A major issue in EC declarations has been determining the boundaries of exceptional droughts, in that the impacts of drought seldom align precisely with the boundaries of Local Government Areas, for example. Brook *et al* (1996) have applied their grasslands model to estimate changes in total standing dry matter across Queensland and also the spatial extent of drought. This approach, which has been extensively tested against field and remotely sensed data, has been of considerable assistance to the Commonwealth Government in evaluating the severity of drought across that State.

Remote sensing offers a unique view into the spatial and temporal variability of vegetation condition and moisture availability (Bullen 1993; McVicar and Jupp 1998). It is consequently being used to help assess which areas qualify for EC specific applications at the Federal level in the drought assessment process, including whether there has been any deterioration in an area since the last visit by government-appointed assessors, to estimate the spatial extent of drought, to determine how much earlier local grasslands have been drying off within a particular year, and to attempt to place current satellite images in historical context. A major value of remote sensing is in the spatial and temporal validation of agronomic models. Improvements in terms of access and analysis to remotely sensed data sets will enable relevant monitoring systems to be operationalised within the foreseeable future.

Stafford Smith and McKeon (1998) highlighted that in order to constrain government expenditure on EC support, it is important to carefully determine appropriate triggers for drought revocation as well as drought declaration. They also investigated the consequences of applying a system that made adjustments over time as to what constituted an exceptional drought through making allowance for ongoing climate change.

Commonwealth assessment of severe droughts now places greater emphasis on the income component of Exceptional Circumstances than was the case in the early years of the policy. In the early years, a threshold condition for the declaration of drought exceptional circumstances (DEC) was that the specified area had experienced a one in 20 to 25 year "effective" rainfall deficit. The other five criteria were then assessed. Since then, several models have been developed to assist decision making.

In assessing the impact on incomes, the Australian Bureau of Agricultural and Resource Economics (ABARE) and BRS provide expert advice (2003). ABARE provides information regarding the financial position of farmers before the event as well as analyzing the impact of the event on income. BRS provides analysis on the event itself and the impact it has or will have on the region.

BRS has developed an integrated toolkit to facilitate the integration of physical and socioeconomic data sets (Clark and Brinkley 2001). Land and Water Australia and other research and development organizations have funded major initiatives to enable farmers to become more self-reliant. In addition to these decision support tools, they have facilitated the development of improved seasonal forecasting throughout Australia (Stone and de Hoedt 2000), and whether or not and how such information can be of value at the farm level (Bowman *et al* 1995; Hammer *et al* 1996; Stafford Smith, Buxton *et al* 2000). Other research in the area addresses the breeding of drought-tolerant plants and improved management of the land (Buxton *et al* 1995; White *et al* 1999a).

It is important to appreciate that a process that achieves reliable assessment of droughts may well be unacceptable to many rural communities. This is because some areas will inevitably receive less assistance than in the past. This is particularly the case in the more arid areas, many of which have received very high levels of government assistance over the past 100 years. Considerable political pressure is likely to be brought to bear to denigrate the assessment process and move EC assessment away from objective scientific measures and back into the political arena.

A perceived deficiency of the process has been the so-called 'lines on maps'. This is in part because the process calls for droughts to be declared within administrative boundaries, even though these seldom coincide with the boundaries of drought. It is therefore not uncommon for adjacent properties to be declared differently with respect to whether or not they qualify for EC, even though to the casual viewer and the media they have been identically impacted by drought. The 'lines on maps' problem has not been as acute in the 2001-03 drought as it was during the 1990s because of the introduction of the concept of 'buffer zones'. While easing any political heat from the debate over EC declarations, this was at best a temporary solution—recognising the inequities generated by lines on maps without attempting seriously to introduce an alternative approach. If differential farm-based support is to be maintained as a policy outcome, a better option is arguably to develop boundaries between regions which make sense in a biophysical sense and then base EC declarations on these zones.

The focus in recent years at the Commonwealth level has been on creating a toolkit to allow biophysical and socioeconomic data for different regions to be more easily and usefully integrated (Clark *et al* 2000). More work needs to be done on testing and applying spatial/temporal agronomic models such as *AussieGRASS* (Carter *et al* 2000) over most of the agricultural areas of Australia. This is because these implicitly incorporate important interactions between rainfall, evaporation and temperature in determining plant growth (White *et al* 1998), take account of differences in soil and plant types, do away with the need to make unreliable estimates of when periods of growth actually occur (these varying considerably with sites and seasons), and highlight the significance of grazing pressure in determining the frequency and duration of droughts (Fouché *et al* 1985), thereby improving the reliability and reducing public controversy over drought declarations.

In terms of the science of monitoring and assessing drought, Australia is undoubtedly a world leader, notwithstanding major undertakings in southern Africa (de Jager *et al* 1998; du Pisani *et al* 1998) and North America (Guttman 1999). The challenge is to ensure that we have the best possible systems in place, and to persuade the community and the political parties that an objective approach for determining when government intervention is appropriate is in the rural and national interest.

In summary, assertions that it is not feasible to objectively measure the extent and severity of drought are not well based and are often made to discredit such approaches for the purposes of political and financial gain. Nevertheless, there are certainly many opportunities for improving both the science and any declaration process that should be capitalized on.

4. The political context

Lindesay has demonstrated in her chapter (Chapter 2) that drought is a normal part of the Australian climate and yet certain sections of the community, including the farming community, continue to react to the onset of drought with surprise and dismay. The media, politicians and the general public often employ the language of conflict when discussing drought—it is our "greatest enemy" and farmers are engaged in a "battle" with it (Wahlquist 2003, p74). This personification of drought is not unique to Australia (for example see Tadesse 2000), but it does reflect a mismatch between Australian expectations and the realities of climate.

An important component of the public and political response to drought events can be explained by a lingering agrarianism in some sections of the Australian community which sees farmers as a 'special' category of business operators and which triggers sympathetic reactions to any hardship they face, often based on little understanding of the realities of Australian agriculture. This section will discuss the importance of agrarian sentiments, both within the farming community and, perhaps more importantly, in the general community and how this translates into political pressure for governments to 'do something' in the event of drought.

Despite the fact that the Australian economy has for the past century been dominated by the services and manufacturing sectors and that the rural population has been declining rapidly (to the point where only around 17% of the Australian population now lives in rural areas), agrarian sentiments about the value of farming as an undertaking have an important place in national sentiments. Described in Australia as "countrymindedness" (Aitkin 1985), these sentiments place a value on farming as an inherently useful and wholesome activity which has distinctly moral overtones. The elements of countrymindedness are very similar to the agrarianism found throughout history and across the developed world (see for example Flinn and Johnson 1974; Mill 1893; Montmarquet 1989; Moyer and Josling 1990, p50). In spite of recent pronouncements by governments about the "transition in outlook from the family farm to the family farm business" (Anderson 1997b, p9024), the image of the hard-working family farmer remains an important part of the national self-image (Lockie 2000, p17). Popular television programs draw on this imagery and Australian athletes have entered Olympic stadiums in Drizabones, emblematic of the "bush". Agrarian sentiments about the moral benefits of farming influenced land reforms in New South Wales in the mid-nineteenth century as policy makers set out to create an "industrious yeomanry" which was as much a social ideal as an economic objective (McMichael 1984, p220). The granting of land to returned soldiers through the War Service Settlement schemes ensured that the rural myth in Australia became tied up with the ANZAC legend. This attachment is as much an urban phenomenon as a rural one.

The existence of agrarianism among many Australian farmers is not to suggest that Australian agriculture is not highly productive, innovative and efficient. Like their counterparts elsewhere in the developed world, Australia's farmers have faced ongoing declining farm terms of trade, and this has resulted in considerable adjustment pressures and impressive productivity improvements. Where other governments have sought to protect their farmers from these pressures, Australian governments since the 1970s have reduced protection dramatically and adopted policy settings designed to support the adjustment process. This has been a necessity given the export orientation of the sector and the limited capacity of a country of barely 20 million to provide substantial domestic support to its farmers.

In spite of the pressures of structural adjustment, the family farm remains the backbone of Australian agriculture and the public perception of the farmer remains that of the rugged individualist. As Wahlquist has pointed out, urban Australians base their understanding of farming on the media and although "Australian farmers are among the most efficient—and least protected—in the world ... nowhere in popular culture is this portrayed" (Wahlquist 2003, p69). Gray and Phillips observe that "more than ever" the stereotypical image of the farmer "appears to be sustained by an urban romanticizing of farm and country life" (Gray and Phillips 2001, p59). The importance of this lingering agrarianism in urban Australia is that the political clout for supporting drought-affected farmers comes not just from the dwindling proportion of voters engaged in agriculture. The comments of urban talk-back radio hosts stir city voters to pressure the government to act—even if only indirectly through their generosity to such charities as the Farm Hand appeals of 1994 and 2002. As has been noted elsewhere, in 1994 "Farm Hand gave the impression that others were stepping in to assist where the Government would not" (Botterill 2003d, p68), albeit that this impression was quite misplaced. Even before the introduction of the Drought Relief Payment in 1994, there had been a considerable relaxation of the access rules to the general welfare system. Among farmers themselves there is a mixed reaction to this support. For every farmer who accesses government drought relief there is a number who neither need nor seek government assistance. The more successful risk managers, however, are reluctant to condemn publicly those farmers who do seek support, although privately many believe that good managers do not require drought support (Wahlquist 2003).

The significance of agrarian sentiments to drought policy lies in the residual sympathy for farmers in the broader community and the way these sentiments play out in public discourse. The following two clear examples of the way agrarianism influenced discussion of drought were provided during the current (2001-03) drought. The first was the treatment by the national newspaper *The Australian* of the issue of government assistance to farmers:

> In June 2002 an editorial, entitled "Farmers need their stockpile of good luck", reported on the recent good times in agriculture and argued that farmers should take advantage of these good times to "provide a buffer for the inevitable downturn in the cycle". The piece suggested that "taxpayers have coughed up election-related sweeteners like the $1.9 billion being doled out to diary farmers ... and the $810 million Agriculture Advancing Australia package that was stitched together to prevent rural voters deserting the coalition". A mere four months later, as a promoter of the Farm Hand Appeal, the editorial in the same paper was opening with a reference to Dorothea McKellar and concluding with the words "this burning continent". The title of the editorial on this occasion: "Lending hand to farmers benefits us all" (Botterill 2003d, p61).

The second example is drawn from Parliamentary debates over drought. In an extraordinary statement, the Deputy Prime Minister John Anderson argued that

> At the outset, I would have to say that this, in many ways, for tens of thousands of farmers and the communities that depend on them, is the cruellest drought of all. We have, over the last couple of years, seen one of the strongest recoveries in rural fortunes of the last few decades. Prices have been very strong. Costs have been down ... But now we have hit this very unfortunate brick wall in most of Australia, where seasonal conditions are really putting that recovery at risk to the point where it is very likely to show up in national economic performance terms (Anderson 2002)

The nature of the family farm, with the intermingling of business and household objectives, complicates the delivery of appropriate drought relief. Until 1994 the NDP provided business support with no welfare component until 1994. This was because policy makers were concerned that the delivery of welfare relief can act as a subsidy to otherwise unviable businesses which will slow the ongoing structural adjustment process. As discussed in Chapter 4, a responsive safety net could ease substantially the pressure brought to bear on governments to act during drought as it is arguably the welfare dimension of the drought impact which resonates most effectively with urban Australia. On the other hand, a safety net could also undermine the objective of self-reliance, as argued in Chapter 10.

Commonwealth and State governments agreed with the NDP against the background of a deep-rooted sympathy for the plight of farmers, as evidenced by public generosity to the Farm Hand appeals of both 1994 and 2002. This residual agrarianism alone, however, would have been insufficient to result in the politicisation of the drought declaration process. Despite the original hopes that it would assist depoliticisation of the drought issue, the current process of declaring the existence of exceptional circumstances drought itself has invited politics into the process at a number of points. Farmers experiencing drought are required to make a case that the dry spell is more severe than could reasonably be encompassed within a risk management strategy. This case is then presented to the Commonwealth Government via the State government. It is also reviewed by the Government's advisory body, the National Rural Advisory Council, which recommends to the relevant Minister whether exceptional circumstances exist. Every step of this process invites publicity and political point scoring. Once the case is in the political arena and public pressure is brought to bear, chances for a science-based outcome which is not influenced by political considerations are often limited.

5. Resolving the tension between the science and the politics of drought

The need to trigger exceptional circumstances welfare support is the primary cause and opportunity for political interference in the declaration process. Where drought-related welfare support is retained, an objective and effective trigger is required if the integrity of the NDP principles are to have meaning. However, recent history suggests that such an option is unlikely to be viable unless there is to be a renewal of political will to implement the policy to which both major parties have said they are committed.

In a liberal democracy with a free press and vocal interest groups, politics will rarely be far from the policy process, particularly when the government wants to be seen to be responding to a group in the community which strikes a sentimental chord. As Heathcote puts it:

> In any catastrophe, public sympathy goes out to the victims, but when those victims are the sons of the soil, on the margins of the good earth, struggling to give us our daily bread, the emotional response is tremendous and objectivity is often left behind (Heathcote 1973, p36).

Drought policy is arguably also about more than objectivity. In particular, it has tended to includes concerns about equity. Based on his modelling work examining criteria for declaring extreme events, Stafford Smith has argued that "as science cannot resolve the balance between objectivity and equity, judgments must be made by policy-makers that inevitably introduce a degree of subjectivity into the process of defining drought declaration and revocation criteria" (Stafford Smith 2003a, p141). This is not to say that an objective system cannot work, but that interpretation of the information requires assessment by analysts who are highly skilled and knowledgeable in assessing agricultural systems. It also requires recognition by policy makers that a drought declaration process which provides many opportunities for politicisation may not result in good outcomes. The challenge is to develop a science-based approach to assessing agronomic drought which is less open to political lobbying at various stages of the process.

Any objective approach to drought assessment, policy development and government intervention requires scientific input. First, scientists have a role in providing expert input into the policy process to ensure that policy makers have an accurate understanding of the constraints of Australia's climate and the impact of drought — thereby ensuring that sound policy goals are developed and met. In order to address the equity dimension, this scientific advice should include input from both physical and social scientists. Second, science can support the means for achieving these ends through, for example, improved risk management by farm operators which, over time, should diminish the political pressure to implement *ad hoc* and counterproductive policies in the face of drought (O'Meagher *et al* 2000).

Australia in recent years has seen a shift in the focus of the government response from support for the farm business through interest rate subsidies to a stronger emphasis on the welfare dimension of exceptional circumstances policy (Botterill, this volume, and Botterill 2003c). This is effective politics in that it addresses the perceived welfare problems which led in 1994 to the addition of the Drought Relief Payment to the original National Drought Policy.

6. Future options

An attractive approach to addressing the politicisation of drought policy is to end the system of drought declarations altogether. A policy approach which recognizes drought as a continuum that encompasses preparation and recovery as well as the drought period itself would be more in tune with the realities of climate. Government support could then be focused on drought preparedness instruments such as Farm Management Deposits. Specifying drought as an "event" rather than part of the climate cycle is problematic in a policy environment based on risk management. The expectation that a drought event will trigger government assistance in some form has the potential to distort the farmer's decision making process and lead to sub-optimal management outcomes.

However, there are significant obstacles to the ending of drought declarations altogether. First, several State governments have their own declaration processes which trigger State drought relief programs. Second, until there is a more mature understanding of Australia's climate among the broader population, politicians and the media, there will be considerable pressure on political decision makers to give recognition to the drought conditions and associated hardship being experienced by farmers. When Prime Minister Paul Keating stated in 1994 that drought was a normal part of the Australian farmer's operating environment, he was lambasted by the then Opposition for "his callous disregard of the rural sector and for his chronic failure to recognize the devastating personal and financial heartbreak of this exceptional drought" (Brownhill 1994b, p34). This was in spite of the apparently bipartisan nature of the National Drought Policy, the principles of which were consistent with the recommendations of the Senate Inquiry into drought policy in 1992 (Senate Standing Committee on Rural and Regional Affairs 1992). Political point scoring has been an unfortunate feature of the implementation of the NDP. A policy which did not provide for recognition of the existence of severe drought conditions through some type of declaration process is likely to trigger the sort of response Keating received when he did little more than restate the basis of the NDP.

A compromise position is to reduce the opportunities for political agitation during the drought declaration process. At present there are a series of opportunities for politics to intervene in the drought declaration process. These opportunities and the problems they raise are:

1) *Requiring farmers to make a case that they are experiencing exceptional drought*: Not all farmers or regions have an equal capacity to present their situation in a manner which will represent their circumstances accurately. This raises the problem of effective advocacy being more likely to attract an EC declaration than objectively severe conditions.
2) *Negotiations between the State and Commonwealth governments*: In Australia's recent political history there have been very few periods in which the State government seeking drought relief has been from the same political party as the Commonwealth government providing the lion's share of the support. This guarantees that both levels of government will use the issue for party political purposes (see for example Amery 2002; Truss 2002).
3) *Commonwealth-State funding arrangements*: In spite of the attempt in the 1992 policy to spell out clearly the funding responsibilities of the State and Commonwealth governments in the event of an exceptional circumstance, the issue of responsibility for funding EC support has been an ongoing point of debate, aggravated by the political realities of point 2.

Australia's Federal system is an important political obstacle to achieving a sustainable, consistent drought policy. Apart from the problems related to the delivery of drought relief itself, the timing of Federal and State elections means that one or other parties to the process is likely to be facing an imminent poll. In this regard, it is of particular concern that the next review of the National Drought Policy, announced by Commonwealth and State Ministers in October 2003 (Truss 2003b), is to be held during a Federal election year.

One option for depoliticising and constraining the abuses of the drought declaration process would be to refer it to an independent statutory body. Examples of such bodies in Australia include the Grants Commission and the independence the Reserve Bank exercises in the implementation of monetary policy. Ideally the establishment of such a body would be bipartisan and enjoy the support of state and territory governments. Though requiring an act of considerable political will, this option is nevertheless feasible notwithstanding recent developments.

An example from South Africa is the National Drought Committee (NDC), comprising national representatives of the Departments of Agriculture, the Agricultural Credit Board, the Soil Conservation Board, the South African Agricultural Union and various agricultural commodity organizations such as the Meat and Wool Boards. The NDC made recommendations on assistance to the Minister of Agriculture, submissions having been lodged by farmers through District Drought Committees to the NDC (O'Meagher *et al* 1998).

Other possible approaches might be along the lines of providing financial support for farm businesses under duress through Government-backed revenue-contingent loans. A scheme of this nature provides default protected access to finance which is only repayable if and when the farm revenues have recovered (Botterill and Chapman 2002). This would be similar in approach to the Australian Higher Education Contributions Scheme (HECS) accessed by tertiary students throughout Australia. As with other approaches, it is extremely important that sufficient rigor be applied to its administration to avoid such access to Federal funding being unduly abused.

An alternative to the above approaches that involves only minor changes to the present system would be to establish a continuous monitoring process comprising officials from all levels of government. Politicians (possibly in the form of the Commonwealth, State and Territory Council of Ministers of Agriculture) could agree only to act on the advice of a semi-independent group of officials with scientific and other required expertise. Drought monitoring would be continuous and advice that DEC existed would be automatically provided once the relevant criteria were satisfied.

Such options would need to be supplemented by ongoing drought-related research and public awareness raising, and could seek to involve farmers both as a way of engaging their support and developing their capacity to provide valuable input in the form of localized knowledge and to learn from the experiences of others.

Addressing the problems which arise at the intersection between science and politics requires a degree of political will and, in the Australian context, bipartisan agreement that once the policy parameters for support are in place, they will not be altered in response to political lobbying. In the absence of such an agreement, farmers and their advocates will continue to exploit the opportunities offered by a democratic system to agitate for the maximum benefit they can achieve. In order to minimize the opportunities for such exploitation, we have argued that the focus of exceptional circumstances support should be on farming as a business enterprise, with the threshold criterion for support being the existence of agronomic drought. Achieving agreement on these two basic points would be a good start to addressing the problems which have arisen in Australia's drought policy in recent years.

CHAPTER 8: DROUGHT RISK AS A NEGOTIATED CONSTRUCT

PETER HAYMAN
Climate Variability, South Australian Research and Development Institute, GPO Box 397, Adelaide, South Australia 5001, Australia

PETER COX*
formerly with Agriculture and Natural Research Management for South East Asia, Catholic Relief Services

1. Introduction

Although agriculture developed in part to reduce the risks associated with hunting and gathering, new risks emerged, especially climate risk (Hardaker *et al* 1997). Climate is clearly a risk to agriculture in Australia where the European settlers struggled to develop farming systems in this strange land with poorly defined seasons. They must have been relieved initially not to have to deal with the winter freeze; the challenge that rapidly emerged was the non-seasonal year-to-year variability, much of it due to the El Niño Southern Oscillation (ENSO) cycle. This difference is evident in the regular breeding cycle of animals and flowering of plants in the northern hemisphere compared with the erratic opportunistic reproduction in Australian fauna and flora (Flannery 1994; Nicholls and Sellers 1991). One of the clearest examples of the contrast in challenges from the climate is the mid-winter feast which was Christianised to become Christmas. In the middle of the European winter, a lot of food is consumed; the notion of a mid-drought feast in Australia would be ludicrous. In the variable climate of Australia, agriculture may be more risky than nomadic hunting and gathering (Flannery 1994, pp280-284).

Drought is a recurring theme in Australian history (Nicholls 1997a) and a feature that distinguished Australia from England. In Britain in 1887, a drought was defined as fifteen consecutive days each with less than one point of rain (0.25mm) (Heathcote 1973). The NSW government astronomer, HC Russell, drew attention to a colonial Australian concept of drought:

> The word drought is not used here as in the sense in which it is often used in England and elsewhere, that is, signifying a period of a few days or weeks in which not a drop of rain falls, but it is used to signify a period of months or years when the country gets burnt up, grass and water disappear, crops become worthless and sheep and cattle die (Foley 1957).

The concept of drought risk in Australia would be simpler if the understanding of drought was restricted to a meteorological term as in the nineteenth Century British

---

*Peter Cox passed away while this book was being prepared.

L.C. Botterill and D.A. Wilhite (eds.), From Disaster Response to Risk Management, 113–126.
© 2005 Springer. Printed in the Netherlands.

definition. The challenge is greater than just developing Australian meteorological definitions of drought that measure time in months rather than days. Russell's definition of drought emphasised the consequences of the rainfall deficiency rather than the exact deficiency. Advances have been made in the development and application of biophysical models to simulate agricultural and hydrological drought (in the words of Russell: disappearing grass and water, worthless crops and dying livestock). Despite the contribution of simulation models, the notion of drought risk remains contentious.

Drought risk is not an extreme irreversible event like global warming and the key management decisions (de-stocking or modifying crop area and inputs) are at a local level rather than requiring intervention at a regional or national scale (compared with the introduction of an exotic pest that can spread rapidly through a wide area). Nevertheless, the effects of drought are usually correlated spatially. This is used as an argument for political intervention. Because many people are affected by drought simultaneously, it is a matter of public concern. Drought is only a source of risk because of the way it interacts with man-made systems such as agriculture and the way it impinges on the way people see, and act in, the world.

In this chapter we examine the risks associated with periodic drought, how farmers perceive them, how scientists try to quantify them, how quantification can be used to improve risk management, and some of the pitfalls in relying too much on 'scientific' (objective) models for risk assessment. Mismatches are noted between different perspectives of drought as risk. These mismatches suggest that the issue of drought is still incompletely specified, that its complexity will always allow multiple interpretations. Useful policy intervention has to respond to these differences and build on them. Just as farming (like science and economics) is socially constructed, the riskiness of farming is a negotiated construct that cannot be understood independent of our minds and culture.

Most farmers in Australia would be surprised that there was much to be written about their perception of drought risk beyond stating the obvious. Indeed, droughts are a tangible reality compared to unseen forces behind interest rates, commodity prices and land values. Droughts are milestones in the lives of rural people. Rural people will often talk about events prior to or after 'the drought' of the 1980s or 1990s. In eastern Australia, years such as 2002, 1994, 1982, 1977, and 1965 spark immediate recognition. These events are remembered because drought, unlike bushfires and floods, is a relentless slowly developing phenomenon that results in a series of stressful decisions, business failures, and neighbours (even partners) leaving never to return.

Farmers with whom we discussed this chapter tended to dismiss worrying too much about words on drought risk. However, a linguist (Arthur 1999) studying drought noted that the Australian language is set to a default country, which is narrow, green, hilly and wet, where rivers run, lakes are full and drought is an exception. The view of drought as an exception makes 'Australia a place perennially disrupted, disjointed and the experience of Australia must always be one of disappointment and suffering'. Heathcote (1973) noted the indignant surprise with which each drought was treated and suggested the cause was part psychological, part patriotic.

2. Social constructs of drought risk—what's so special about drought?

In 1989 Kraft and Piggott (1989) posed the question as agricultural economists, Why single out drought? and developed a coherent argument that it was problematic to single out one source of income fluctuation in one sector of the economy for special treatment. This question was repeated by agricultural economists a decade later (Thompson and Powell 1998). Farmers are well aware that slumps in commodity prices can have greater impact on farm well-being than drought. Yet within rural communities, the media, the wider community and government, drought is perceived as a special risk. Each culture accepts some risks as normal and views others as noteworthy (Douglas and Wildavsky 1982). An example is the risk of car accidents compared to train or airplane accidents. Climate may be viewed as special because over most of history, climate was understood to be controlled supernaturally and extreme events taken as a sign of God's wrath. This contrasts to the neoclassical economic treatment of climate variability in which a perfectly informed society adopts optimal strategies, yet the number of farmers who have attended prayer meetings for rain possibly equals attendance at scientific workshops on El Niño. Indeed, El Niño is regularly personified in cartoons and press stories. Jill Ker Conway (1989), in her autobiographical account of the drought in the 1940s in which her family lost the farm and her father died probably as a result of suicide, observed that the drought caused her 'to lose her faith in a benign providence' and as a young girl to take to heart Shakespeare's King Lear: 'As flies to wanton boys are we to the gods. They kill us for their sport.' (Shakespeare 1996, Scene 1, Lines 36-37). This sentiment resonates with Henry Lawson's poem Beaten Back:

> Can it be my reasons rocking,
> for I feel a burning hate
> For the God, who only mocking –
> sent the prayed for rains too late.
> (Lawson 1988, p52)

Across the wider community, drought receives special treatment. In a commentary of how the media and government treat unemployment and drought Watts, (2003) noted that both are associated with psychological harm, ill health, poverty and social exclusion and both have elements of a structural or systemic failure on one hand and personal responsibility on the other. In the case of unemployment, the emphasis was on personal responsibility while the structural or systemic failure is downplayed. This was contrasted with how farmers are treated as victims of a systemic problem of drought with limited emphasis on the role of their management.

The Farmhand appeal was launched in October 2002 by prominent businessmen and media personalities to assist drought-affected farmers. A significant portion of the 22.5 million raised by Farmhand came from small individual donations and along with other funds raised by churches and charities reflected a sense from urban Australia of wanting to connect with rural communities and/or a belief that drought-affected farmers were the deserving poor. This treatment of drought risk as special has strong agrarian or agricultural fundamentalist roots, with the basic notion that the family farm producing food and fibre has an inherent worth and should be protected for the good of the nation.

If drought policy is based on changing how farmers manage the risks associated with periodic drought, we need to understand underlying cultural perspectives that colour the way in which drought is perceived as a risk and hence how the changed signals will be interpreted. In the following section, support for drought assistance can be summarised as arguments about the symbolic role of agriculture, an efficiency argument and a welfare argument. A final argument revolves around issues of animal welfare and land degradation.

2.1 VIEWING DROUGHT AS A SYMBOLIC THREAT TO THE AUSTRALIAN COMMUNITY

An analysis of hundreds of Australian media articles, parliamentary speeches, books, poetry and films suggested that drought was regularly invoked as a symbolic threat to the Australian national community (West and Smith 1996). Furthermore, this symbolic use of drought showed no sign of waning, despite the declining relative importance of agriculture to the national economy. In another analysis of the wording of newspaper reports of droughts from 1900 to 1995, little had changed (Arthur 1999). Droughts are declared; drought is something we must combat, and battle with a plan of attack. Drought grips, creeps, bites and decimates the land and people who are drought-smitten, desperate, ruined. Land that is irrigated or in higher rainfall areas is called safe country. Safe from the 'dread enemy' as drought was called in 1906, or 'our biggest enemy' in a headline in 1995. If drought is a similar threat to war the nation is at risk and government intervention is easily justified. In the context of the 1994 drought, a senator described drought as a time for all levels of government and all sides of politics to work together through a national tragedy (Brownhill 1994a, p1681). In December 2002 John Ubergang, a farmer from northern NSW, proposed the Crooble Plan (Ubergang 2002) to avoid the 'major social and economic chaos and disaster for Australian rural producers and small businesses'. He proposed that the Federal Government should take 3% of the national budget to fund a national disaster management program which would be allocated by a team of local farmers, graziers, agronomists and veterinary surgeons once any local weather station recorded a two-monthly rainfall below 50% of the long-term average for those two months.

2.2 VIEWING DROUGHT AS A RISK TO THE WELFARE OF RURAL FAMILIES AND COMMUNITIES

Drought is recognised as a factor in divorce, suicide and illness in rural areas (Munro and Lembit 1997). Although not the exclusive cause, drought serves as a catalyst for major upheavals in rural communities; it is the focal point for structural problems of farm size, falling terms of trade and the fragile interdependence of rural communities. To those feeling these compounding effects, drought is an intense lived experience rather than part of a probability distribution or risk profile. In the context of researching lived experience, Virginia Woolf's description of a metaphor has been used as a means of not describing the object itself but providing 'the reverberation and reflection...close enough to the original to illustrate it, remote enough to heighten, enlarge and make splendid' (Van Manen 1990). In this sense, drought is a metaphor for rural hardship and suffering.

When then Prime Minister Paul Keating pointed out in 1994 that drought and climatic variability are part of the natural environment and did not constitute a natural disaster, he was following a line that de-mystified drought. At the time, he was criticised by the media for a callous comment. An indignant opposition senator claimed he was 'condemned all over Australia by every person in the rural community for saying that drought is a natural recurring phenomenon' (Senator Brownhill cited in West and Smith 1996). Although this was partly party politics, when a policy economist asks a farmer to consider drought as a normal recurring business risk, some interpret this as asking farmers to take the enlarged metaphor of rural suffering as a normal and recurring risk.

2.3 VIEWING DROUGHT AS A RISK TO THE EFFICIENCY OF THE RURAL, AND HENCE NATIONAL, ECONOMY

The underlying notion of keeping the farm going for efficiency rather than welfare reasons is strongly held. A recent example is the Queensland Western Downs Solutions Group formed in March 2002. In their request for a one-off injection of $10 million and reforms of drought assistance, the justification for funds was based on the 'past and future contributions of the drought affected shires to the economy' (Australian Broadcasting Corporation 2002). In this sense, drought is a risk to the economy and drought support is part equity for past contributions and part investment for future contributions. Efficiency and equity in terms of the adjustment process was used by the NSW Farmers Association in their submission to a review of assistance arrangements:

> The well acknowledged rationale for public assistance in this case is that it prevents inefficient adjustment that might otherwise occur. Australian farmers receive very little Government assistance, especially compared to farmers overseas. It is therefore appropriate that they receive assistance to withstand events which may cause them to exit the industry, if they are otherwise viable. (NSW Farmers Association 2001)

Agricultural economists (Freebairn 2002a; Freebairn 1983; Simmons 1993) argue that: a) the underlying assumptions of risk to the breeding stock and skilled labour through a drought are generally overstated; b) drought assistance and transport subsidies can lead to environmental risk by overstocking and keeping stock too long into a drought; and c) assistance leads to moral hazard, that is, a cross subsidisation of the careless by the careful and, in some cases, of the dishonest by the honest. If climate is treated as a unique source of risk requiring assistance, it is likely that there will be too much investment in farming drought-prone regions compared to regions with more reliable rainfall. Furthermore, across all regions, farmers are likely to pay too little attention to managing climate risk compared to other risks if this source of risk has already been hedged by government policy.

2.4 VIEWING DROUGHT AS A RISK TO ANIMAL WELFARE AND THE LAND

Drought is an extremely high-risk decision environment, especially for graziers. As the drought worsens, prices fall and the cost of feeding increases. The decision to sell is made more difficult by concern that prices might rise when widespread rain comes. The advice in the past could be summarised as either 'sell and regret, or let the worthless animals die' (Anderson 1994). The second option is now illegal on animal welfare considerations. In any case, the classic drought photo of the skull on parched ground represents a situation that should not occur, as stock should be removed long before they starve and/or the groundcover is reduced.

Most rural producers are concerned about the environmental impact of drought. However, it is likely that more and more urban taxpayers are likely to see drought as a risk imposed by people on the environment. There is an increased preparedness to challenge current farming practices (a situation not dissimilar to the public reaction in Britain to BSE and foot and mouth disease where sympathy for farmers was mixed with a challenge to farming practices). Following an article in the Sydney Morning Herald on the drought and the request from farmers for assistance in July 2002, two letters appeared, neither of them sympathetic. The first letter (Cohen 2002) observed:

> Very little time seems to pass between claims that farmers are in trouble and having a hard time. How often have we seen them receive financial assistance because of a drought, flood, loss of stock or crops? Compare that with the number of times we hear about small business being offered money to enable them to continue trading..... If the cockies find life so hard on the land, why don't they leave it and try to make a life in the city, as many of us are trying to achieve?

The second letter (Bensen 2002) expressed concern over the environment:

> Australia is periodically affected by drought due to the cyclical El Niño climatic effect. This has been going on for tens of thousands of years. However, Europeans have been settled here for only 214 years and appear not to have adjusted to this phenomenon.... One could ask when we are going to put limits on intensive agricultural development in this fragile, drought-prone country.

The drought of 2002 has been labelled the first green drought (Megalogenis and Wahlquist 2002) in reference to discussion on appropriate ways of managing Australia's land and water resources. This debate became prominent when some media personalities associated with the Farmhand appeal proposed ways of drought proofing Australia. Media coverage of the debate was maintained by the response by the Wentworth group of scientists, who argued that any attempt to drought proof Australia was foolish. It is likely that urban taxpayers will increasingly join the debate on drought risk between farmers and governments.

3. Different ways of defining risk

Not only is drought a contested term, risk as a term can be used in both a technical-legal sense and in everyday language. Risk is broadly defined here as uncertainty with consequences (Anderson and Dillon 1992). Risk management involves reducing

ing peril and exploiting opportunities (Clark

olicy development include: defining risk as
ı distinction between risk and uncertainty;
n or as multiple interacting causes; the
tribution that provides a description of the
. psychological process.

S RISK AS VARIANCE

ıance of loss (Delbridge *et al* 1991) and this
    Amongst economists, risk refers to the
sults from an action. Risk analysis considers
come and its variance or, in the case of
ıs of distribution of different outcomes. A
down-side risk (negative deviation) is
e (Hardaker *et al* 1997). However, if the
ır example, a normal distribution), it makes
est that the distinction is not required. If
ıe outcomes are catastrophic, use of more
ːh as negative semi-variance or negative
mmetrical distribution of crop yield is likely
ts, which will change again when tax is
ıf outcomes will not matter to risk-neutral
ı interested in maximising mean wealth and
nvolved. Needless to say, most people are
.-side risk as well as long-term averages of
Australian farmers are slightly risk averse
ːr 1980).

The word 'risk' comes from Italian *risicare,* which means to dare and emphasises
choice, opportunity and gain rather than fate and loss (Bernstein 1996). In general,
variability is not a bad thing: it sometimes allows us to recover from our mistakes; it can
be a source of novelty; and it provides a screen against which to choose between
alternative decisions. Furthermore, the ability to handle variability is one of the sources
for private entrepreneurs to gain competitive advantage and succeed (Malcolm 1994).

3.2 RISK VERSUS UNCERTAINTY

Another aspect of risk which has policy implications, especially for the funded research
and development in the wake of the drought policy, is the distinction between risk and
uncertainty introduced in 1921 by Knight. A simple example is considering the toss of
a fair coin as risk whereas the toss of a biased coin is uncertainty. After experimenting
with the biased coin, the uncertainty could be quantified as risk.

This notion underpins much of the role of science assumed in current drought policy. The reasoning is that the future is uncertain, and that, using rainfall records and crop and pasture models, some of this uncertainty can be quantified as risk. Central to this approach is the frequentist view of probability distributions which uses historical data (measured or simulated) to produce a probability distribution of outcomes. The alternative subjectivist view uses probabilities to capture the degree of belief an individual has that a given outcome will occur (Hardaker *et al* 1997). This is not only expedient, but a legitimate way to construct a decision calculus. Decision makers may adopt the output from a scientific model or historical rainfall data, but this then becomes their subjective view.

The collation or modelling of derivatives of rainfall, such as animal or crop production, requires considerable judgement by scientists and trust in that judgement by farmers and advisers. These judgements are usually not quantified and, in a frequentist sense, hardly quantifiable (Matthews 2000). The trust requires an assumption that historical records are still applicable and that the models used reliably transform these data into probabilities of consequences. The contribution of biological scientists is partly based on the idea that it is both possible and desirable to specify objective probability distributions that describe critical system parameters. However, for many farm management problems, this is only partly possible (because the system can only be partially specified in an engineering sense), and may not even be the most effective means of intervening and improving the management of farming systems.

3.3 TREATING ISOLATED SOURCES OF RISK OR MULTIPLE INTERACTING SOURCES

Although textbook treatments of decision making under uncertainty often deal with a single source of risk which has a fixed relationship with the outcome (for example, the impact of seasonal rainfall on farm returns), this is rarely the case in practice. Climatic variability is a major contributor to agricultural production risk, but other factors also influence production risk, such as pests and diseases or responsiveness to fertilisers. From the point of view of farm management, production risk is only important if it affects business risk. The other important component of business risk is price risk. A further complexity is added by consideration of financial risk which is the variability of net returns to owners' equity after financing debt. During the 1990s drought, interest rates for some farmers were as high as 23% per annum. As the vast majority of farms in Australia are family farms, one experienced financial consultant included in drought risk the risk to family labour 'who are underpaid and in many cases overworked and ill rewarded for all their efforts, but trapped into trying to preserve the family capital through desperate measures' (Peart 1992). Others have added the considerable occupational and safety risks of rural work and the risk of divorce. None of these risks are rare, but they are exacerbated by the stress of drought, and are often fatal for the farm business (Anderson 1994).

Consistent with a more holistic view of risk, the adjustments to exceptional circumstances led to tighter mathematical definitions of what was meant by rare events, but broader definitions of factors that could be considered as exceptional circumstances.

This is supported by whole-farm stochastic modelling that showed drought as just one factor causing serious decline in net farm income (Thompson and Powell 1998). However, it has led to some angst amongst farming communities, as more judgement is required and hence the process is perceived to be more readily politicised and bureaucratic (NSW Farmers Association 2001). This introduces policy risk which makes planning even more challenging.

Alternative and sophisticated holistic thinking on drought risk comes from work in developing countries where the collision between drought and poverty is dramatic. One of these approaches, vulnerability analysis, turns conventional impact analysis on its head by considering multiple causes of critical outcomes—dislocation, hunger, famine—rather than the multiple outcomes of a single event such as drought (Blaikie *et al* 1994; Ribot 1996). Another distinction is made between risk as *ex ante* income management and coping with bad outcomes through *ex post* consumption management (Webb *et al* 1992).

In an Australian context, Malcolm argued for a broader view of risk and suggested that the recent enthusiasm for risk management was partly due to misinterpreting problems of low farm income as being due to poor risk management. Poor income is more likely to be due to structural problems at farm and industry levels. Although climatic risk may exacerbate the problem, information and procedures on climate risk will do little to solve the problem if the underlying cause of low income is farm size or inappropriate land use (Malcolm 1994).

3.4 REPRESENTING RISK AS A PROBABILITY DISTRIBUTION OR A PSYCHOLOGICAL PROCESS.

Much recent research into the psychological process of risk perception in agriculture relates to quarantine (Finucane 2000), biotechnology (Coakes and Fisher 2001), and pollution (Hattis 1994). A psychometric model of risk uses scaling techniques to systematically measure responses to a series of hazards. These methods have shown that hazards judged as dreadful and unknown are also judged as the most risky (Finucane 2000). Psychological studies have identified various issues that influence the perception of risk, including the subject's sense of control and her worldview, whether a risk is voluntary, and the distribution of costs and benefits. Feelings about, or response to, risks are central to a lifetime of learning (Damasio 1994): the point about learning is finding out progressively what is and what is not possible, and using this knowledge to change behaviour—that is, manage risk.

4. Do farmers underestimate the likelihood of drought?

Most, if not all, approaches to drought risk have some notion of the likelihood of low rainfall. A view of drought as a mismatch between farmers' expectations of rainfall and the rainfall that occurs implies that farmers tend to be over-optimistic and misjudge the variability of the climate. Supply of objective rainfall probability data by researchers is held to correct this misperception. The argument for Australian farmers misjudging the variable climate comes from the muddle of past drought declarations, historical analysis of a pioneering optimistic sprit, literature from the field of psychology and limited empirical evidence.

4.1 THE MUDDLE OF PAST DROUGHT POLICIES

The history of official drought declarations suggests the refusal to accept the reality of the arid and variable climate—a reality that it is in abnormal, rather than normal, seasons that Australia is green like England. In Queensland, some shires have been either partially or completely drought declared 70% of the time from 1964 to the early 1990s. In NSW, some districts have been drought-declared for three months or more for 65% of the time (Simmons 1993). There is a striking difference between these frequencies of drought declarations and the meteorological definition of drought as the driest 10% of years in the historical record (Gibbs and Maher 1957), or the definition of exceptional circumstances as the driest 5% of rainfall for a defined period. Some care must be taken in using drought declaration for evidence of misjudging climate as it might be just effective advocacy. Recalling his days as Finance Minister in the 1980s, Peter Walsh (1994) referred to drought declaration as a regional licence to milk the Canberra cash cow. Some of the inconsistencies between states have been analysed (Smith *et al* 1992) and support Walsh's assertion.

4.2 THE GOYDER LINE AS A WARNING TO AGRICULTURAL OPTIMISM

The 'Goyder line' was named after the first Surveyor General of South Australia who, in 1865, established a line to mark the limit that drought had extended south. This was done to assist in determining which pastoral leases might be given financial relief (Andrews 1966). It became a line beyond which cropping should not proceed. It was ignored during a run of good seasons, with dire consequences. Governor Phillip's response to the drought in 1791 in the early days of the colony was that he did not think it probable that so dry a season often occurred. This optimistic view of climate has characterised much of the pioneering spirit in Australia (Nicholls 1997a). The optimism bias is recognised in psychology (Weinstein and Nicolich 1993); the optimistic view that bad things are more likely to happen to other people than to us is one of the historical reasons for legislation that employers, not employees, take on insurance for accidents to workers (Moss 2002).

4.3 EVIDENCE FROM PSYCHOLOGY THAT HUMANS FIND IT HARD TO THINK ABOUT RISK AND UNCERTAINTY

People (not just farmers) are poor intuitive statisticians. And most of us refuse to admit it. Shanteau (1992) noted the widespread evidence (including studies with farmers) that people have limited cognitive capacity to process low probabilities and focus more on the magnitude of single outcomes than the likelihood of different outcomes (Table 1). These biases have been related to people's understanding of climate variability (Nicholls 1999; White 2000a).

Most of these biases relate to imperfect sampling, whereby we are prone to use the wrong distribution to derive the likelihood of an event, and even if we have lived long enough our memories can't help but be coloured by emotions associated with events such as droughts and good seasons. Although the effect (feeling) heuristic can be efficient, it can also mislead (Finucane et al 2003). On this basis, science with easy and dispassionate access to long-term climate records appears to have the upper hand on farmers who have unreliable memories of limited time series. However, this argument must be used with caution. Because of climate change the underlying sample may be changing; there are clear shifts in temperature, evaporation and to a lesser extent rainfall (Hennessy *et al* 1999). Furthermore the sample of wheat yields or carrying capacity may not be all that simple and straightforward because of technology trends and complex interactions with degradation events. Some of these changes can be captured in simulation models but others cannot. A counter argument to the gambler's fallacy for climate is the notion that the oceans, the engines of periodic drought, unlike a coin, do have memory. In any case, some agricultural processes are self-correcting such as putting on fertiliser which is not used for a crop but is available for the following season.

Table 1. Perceptual biases (adapted from McCall and Kaplan 1990; Nicholls 1999)

Bias	Description
Availability	Availability bias means that an estimate of the frequency of an event can be influenced by the vividness of the imagery. Items or events that are vivid or presented first (primacy) or last (recency) assume undue importance. The proximity of a drought or a run of good seasons is likely to affect the rated frequency.
Selective perception	People seek information consistent with their own views. Many farmers have strong views on cycles of floods and droughts and look for confirming evidence. Scientists look for confirming evidence of the value of seasonal climate forecasts.
Concrete information	Vivid, direct experiences dominate abstract information; a single personal experience can outweigh more valid statistical information. Farmers are likely to remember the booms and busts more than the average years.
Law of small numbers	Small samples which are readily available to a decision maker are seen as representative of the larger population even when they are not.
Insensitivity to base rates	In dealing with uncertainty, people often ignore background information; for example, that there are about twice as many El Niño events as there are bad droughts in a given location in eastern Australia.
Gambler's fallacy	People can be convinced of patterns that don't exist; for example, that random events are self-correcting. By the autumn of 1994, farmers in southern Queensland had had three bad seasons; when it rained in autumn they believed that they were due a good season, only to get the worst El Niño drought since 1982.

In a previous text on drought policy in 1973, Heathcote hypothesised that 'folk' perceptions of drought risk are much lower than the 'scientific' assessment of drought risk. There is some support from a study in Western Australia in 1984, which found that drought frequency was underestimated and viewed incorrectly as having a regular cycle rather than being episodic (Conacher and Conacher 1995).

A different finding to the farmer as a misguided optimist emerged from interviews with 90 farmers and 20 advisers in northern NSW to document their subjective risk assessment of seasonal rainfall and derivatives of rainfall such as fallow recharge and crop yields (Hayman 2001). When these subjective risk assessments were compared to the long-term rainfall record and simulated yields and fallow recharge, farmers saw the climate as drier and more risky than the long-term record suggests; they rated the chance of crop failure and low yields much higher than crop simulation models did. This contrasts with conventional wisdom which holds that farmers are overly optimistic and need access to the long-term rainfall record to appreciate the true risks of farming. Furthermore, most farmers interviewed had lived through a wetting trend. A farmer with perfect memory of only the last 20 or 30 years might be expected to view the climate as wetter than the 100-year record suggests. Such evidence as we do have suggests that farmers may be too conservative in the allowance they make for the possibility of drought. In any case, farmers may be able to adapt quickly to the different set of outcomes engendered under the National Drought Policy, with or without access to additional quantification or transformation of rainfall data.

5. Different models of managing risk

From a psychological perspective, Hammond (1996) argued that uncertainty in the world outside the observer generates uncertainty in the observer's cognitive system. This view allows for a continuum in the uncertainty of the external environment from a highly controlled environment (for example, a chemical engineering plant) to a natural environment. As the discipline of agricultural science shifts from reductionist science to the study of farming systems and then to applied ecology, the quantification of uncertainty as risk becomes more difficult. Systems are less well specified; there are far more linkages including feed-back and feed-forward loops; and we are dealing with competing claims about what will happen and what should happen. Hammond's systems view of risk argues that cognitive systems (the way we order experience and construct reality) also have a continuum from intuition, which takes little account of uncertainty, to analytical thought which in extreme forms such as decision analysis attempts to account formally for uncertainty.

There is a mismatch between the understanding of risk by farmers dealing with complex systems with clever but relatively simple intuition (Cox 1996; Cox *et al* 1995; Gigerenzer and Todd 1999) and the treatment of risk by scientists modelling relatively constrained agricultural production systems with complex formal analysis. In terms of managing risk for a farm business it is better to solve the whole problem roughly than attempt to solve part of the problem extremely well. Even at the whole-farm level, risk is only a partial issue. Farm management economists have argued that research and development may have swung from ignoring risk to placing too much emphasis on risk (Pannell *et al* 2000).

6. How does agricultural science deal with risk as a negotiated construct?

The National Drought Policy (NDP) of 1992 emphasised farmers' risk management. The primary role of government was switched from one of disaster relief during a drought to one of ensuring that farmers are equipped to manage climatic risk effectively themselves. Research and extension providers were funded to assist this transfer of the responsibility for managing climatic risk from government to farmers. The role of publicly funded research and development in providing information (rainfall probabilities, simulated crop and pasture production data based on historical climate records) and training (workshops, decision support systems) was generally treated with enthusiasm by scientists and economists. One exception was the farm management economist, Bill Malcolm (1992; 1994), who maintained that there were two untested assumptions underpinning much of the research and development in the wake of the NDP: that farmers were poor managers of risk; and that R&D could help them manage risk better. He pointed out that, even if the first claim were true, the second did not necessarily follow. The NDP encouraged communication between farmers and researchers about the risk of drought and the riskiness of different decisions for managing it. But this was not as straightforward as it first appeared.

The process of increased engagement between rural producers and the professional R&D establishment following the NDP did encourage scientists to consider risk in their analyses and develop better ways of engaging with farmers (Nelson *et al* 2002). We have argued that the exact specification of probability distributions may be a more modest contribution than first thought. However, probability distributions of rainfall and production have started to provide a common language and a Rosetta stone for communication about risky decisions. For the biological researcher, it opened up the uncertain decision-making world of practical farming and farmers were introduced to more methodical ways of framing farm management decisions. It helped clarify the thinking of farmers, researchers and policy makers about our environment and issues of resource allocation.

One of the lessons was the challenge of capturing drought risk in a single probability distribution or risk profile. The US Environmental Protection Agency maintained that a hard lesson for technical experts was to acknowledge that, in the process of risk characterisation, apparently inexplicable inconsistencies may be recognised as responsible, reasonable descriptions of the same problem (Patton 1993). When scientists use terms such as 'real risk' and 'perceived risk' to dismiss people's concerns about pesticide use they have been accused of 'unreflective use of ambiguous, emotion-laden words from everyday language' (Beck 1992). That is not to say that the contribution of science to assessing and managing drought risk is trivial; rather, it is partial and will be most useful when it is clearly stated as partial. Although a probability distribution on a computer screen may be a contribution, we must recognise that risk is, and will always remain, a negotiated construct that cannot be measured outside of the mind and culture of the decision maker.

If what we are saying is correct, the marginal returns to more elaborate formal analyses to improve the efficiency of drought management may be modest. The key issues may be elsewhere—finding the best way to encourage farmers and their communities to make their own decisions but provide a welfare safety net; determining how environmental and production values can be managed together during a drought; a switch from analysing micro trade-offs between risk and return within a given enterprise (for example fertiliser rates or crop choice) to encouraging a diversity of responses to drought, including diversification of rural livelihoods (Ellis 2000); or deciding the extent that an urban population values farming communities and their contribution to natural resource management.

CHAPTER 9: PROSPECTS FOR INSURING AGAINST DROUGHT IN AUSTRALIA

GREG HERTZLER[*]

School of Agricultural and Resource Economics, The University of Western Australia,
35 Stirling Highway, Crawley WA 6907, Australia

1. Introduction

For more than a century, countries around the world have implemented crop insurance programs (Hazell 1992). Most of these programs insure against multiple perils, including drought. For more than a century, these programs have failed. None has been commercially viable. All have been subsidised and many have become too expensive for governments to afford. In general, countries that continue to subsidise agriculture also continue to subsidise their crop insurance programs. Canada's crop insurance program has loss ratios up to 3 (Sigurdson and Sin 1994). The indemnities paid out by insurers plus the administration costs of the program are three times greater than the premiums paid in by farmers. The US program has similar loss ratios (Gardner 1994) and currently subsidises 67% of the premium for farmers who insure against yields falling below 50% of average (Skees 2001). Brazil and Japan have loss ratios above 4.5 (Hazell 1992). Recently, the Europe Union investigated insurance as it reforms its Common Agriculture Policy (European Commission 1999). Almost unique among its competitors and trading partners, Australia has been unwilling to directly subsidise farm programs, including crop insurance.

In Australia three studies have investigated the viability of multi-peril crop insurance. In 1986, the Industries Assistance Commission recommended against a crop insurance program (Industries Assistance Commission 1996). In 2000, the Multi Peril Crop Insurance Project (Ernst & Young 2000) concluded that crop insurance was not feasible without government subsidy. In 2003, the Multi Peril Crop Insurance Task Force (Multi Peril Crop Insurance Task Force 2003) conducted a detailed analysis for Western Australia, the largest and most reliable wheat producing state in the country. If crop insurance is viable in Australia, it will be in the state of Western Australia. The Task Force, however, 'saw no future for multi-peril crop insurance in the absence of significant government subsidisation of premiums or underwriting of risk'.

Are there any prospects of insuring against drought in Australia? Surprisingly, the answer is yes. Financial markets may succeed where governments have failed. Around the world, methods for financial risk management are being extended into climate risk management. Weather derivatives are offered by global financial institutions to ensure against too much or too little rain, too hot or too cold temperatures. Yield index

---

[*] I would like to thank Jerry Skees and Ben M. Gramig, who reviewed the chapter and helped clarify important points, and Brian Hardaker and Barry White who provided very useful comments and corrections.

L.C. Botterill and D.A. Wilhite (eds.), From Disaster Response to Risk Management, 127–138.
© 2005 *Springer. Printed in the Netherlands.*

contracts are based on weather derivatives as a replacement for crop insurance. Two further recommendations of the Multi Peril Crop Insurance Task Force (2003) were:

- Determine what can be done by government to assist in:
  o   Setting up required infrastructure for weather derivative products;
  o   Developing independent, reliable data collection; and
  o   Improving grower knowledge of the products and their potential value to farmers.
- Consider how government could assist in developing a suitable model on which to base a relevant index for farmers that has a strong relationship to Western Australian crop performance.

Although pilot projects are beginning in a few developing and transitional countries (Skees 1999), Australia is a unique laboratory for experimenting with weather derivatives and yield index insurance. It has a well developed financial sector and any commercially viable scheme will not be crowded out by government subsidies. Subsidies may attract crop insurers to North America, but the benefits of diversifying their portfolios will attract them to Australia.

This chapter reviews markets for risk and why they are needed, crop insurance and why it fails, early proposals for rainfall insurance and why they were never implemented, and current proposals for weather derivatives and yield index contracts and why they might succeed. It concludes with a research agenda to fill in the gaps in our knowledge.

2. Markets for risk

With so many failures over almost a century, why is crop insurance still on the political agenda? An uncharitable answer is that farmers and insurance companies lobby governments for their own advantage at the expense of society as a whole, behaviour that economists call 'rent seeking' (Goodwin and Smith 1995). As subsidies are traded away in the negotiations of the World Trade Organization, other forms of subsidies are implemented. For Australian farmers, it would be hard to view the resurgence of crop insurance in the US in any other way. Although subsidised crop insurance was promoted as a replacement for *ad hoc* disaster aid, disaster aid continues as insurance subsidies increase (Skees 2001). Not surprisingly, some Australian farmers have lobbied governments to 'level the playing field' and introduce multi-peril crop insurance in Australia.

Another answer is that markets are failing to provide a necessary service for farmers and governments should correct the failure. Risk sharing is an essential service provided by financial and insurance markets. Effectively, the risk is transferred to people who are better placed to manage it and the risk is diversified throughout the economy. For their service, people who bear the risk are paid a risk premium. Farmers benefit by more access to available credit, more ability to be entrepreneurial and adopt new technologies and more specialisation and efficiency in production (Arrow 1996; Goodwin and Smith 1995; Skees 1999). Perhaps there is a significant demand for crop insurance but the insurance industry is unable to supply it (Miranda and Glauber 1997). The recent bankruptcies in the Australian insurance industry support this conclusion. Insurance is not like other commodities that are traded in smoothly functioning markets and crop insurance is one of the most difficult of insurance products.

We take markets for granted and call them failures when they don't work. It is easy to forget that markets depend upon ideas and technologies and are, essentially, inventions. A viable market for crop insurance has yet to be invented. Every market must solve the problems of moral hazard, adverse selection and transaction costs. In addition, risk markets must solve the problems of basis risk and systemic risk. Moral hazard is sometimes called hidden action. It becomes a problem if someone is able to subvert the outcome of a trade once the deal has been struck. Contracts are necessary and markets have legal and administrative mechanisms to enforce contracts and verify that people comply. Adverse selection is sometimes called hidden information, information known by one person and not another. Insider trading is illegal because markets are voluntary and people don't volunteer unless they are sure about what has been agreed on. Together, adverse selection and moral hazard are sometimes called asymmetric information. Transaction costs may be higher than the benefits of a trade. This is often true for insurance contracts, especially those that are tailored to individual circumstances. Basis risk occurs for the opposite reason. To keep transaction costs low, insurance and financial markets trade in standardised risk contracts. Automobile and homeowner's insurance are 'much of a muchness'. Futures and option prices are the same for everyone, but these are not the prices a person will actually pay or receive for their commodities or stocks. The difference is the basis which may change unpredictably, causing basis risk. For this reason, basis risk is sometimes called imperfect indemnity. Systemic risk can bankrupt the system. If many people are insured for the same risk such as a natural disaster, a change in the price of wheat or a crop failure, there may not be sufficient capital reserves to make the indemnity payments. Insurance markets deal with systemic risk by avoiding them, keeping capital reserves and reinsuring to diversify the risks throughout the industry. Financial markets are designed to diversify systemic risks over a large volume of traders.

Crop insurance programs have not solved the problems of moral hazard, adverse selection, transaction costs and systemic risk. Financial markets for weather derivatives solve these problems, but not the problem of basis risk. Perhaps the best of both crop insurance and weather derivatives can be combined to create a commercially viable market in which farmers and rural businesses routinely purchase insurance against climate risks.

3. Multi-peril crop insurance

Farmers are willing to buy insurance and insurers are willing to sell it to them. Fire and hail insurance have been available in Australia for some time. Why are these viable if multi-peril crop insurance is not? Fire and hail insurance have low moral hazard. There is very little farmers can do to make it hail, and lightning strikes appear to be the major cause of fires. In addition, damages are easy to assess. Fire and hail insurance have minimal adverse selection. Information is available to both farmers and insurers with good long-term records of damages. The transaction costs were high initially but have become much lower with experience. Systemic risk is low. Fire and hail are independent events that affect only a few farmers at a time and reinsurance is available for local insurers to share the risks with larger insurers. For all these reasons, the premiums for fire and hail insurance are low. In Western Australia they range from 0.5% to 2.5% of the crop's value with an average premium of less than 1% (Multi Peril Crop Insurance Task Force 2003).

For multi-peril crop insurance, moral hazard may be the least of the problems. Even so, there are ways that both farmers and insurers can subvert a crop insurance contract. For example, farmers may fail to fertilise or spray for pests. Or they may not check that the harvester is adjusted and working efficiently. This aspect of moral hazard is usually managed by coverage levels. The insurer will only cover a proportion of production, say 65%. The farmer covers the remaining 35% and still has an incentive to grow a good crop. During a bad season, however, production may surely fall below 65% and the farmer may get paid for 65% of an average year regardless of how poorly the crop yields. To make sure that farmers continue to care for the crop, insurers apply an election percentage of around 70%. In case of a complete crop failure, a farmer will only get 70% of 65% or 45.5% as an indemnity payout. If yields are 50% of normal, a farmer will get 70% of 15% or 10.5% as an indemnity payout plus 50% for selling the crop, giving a total of 60.5%. A farmer could also subvert the contract by selling the crop to a neighbour and filing an insurance claim. To prevent this, insurance adjustors must assess the crop before harvest, a costly and time consuming task that increases the transaction costs. Insurers may also subvert the contract. They may not keep sufficient reserves or not reinsure and be unable to pay the promised indemnity claims. Farmers have little recourse in this circumstance.

Adverse selection is a major reason most crop insurance schemes fail. Farmers know more about their farms than does the insurance company. Verifiable data on farm yields rarely exists. There are two important types of information the farmer knows but the insurer does not. The first type is average yields. Because of data problems, premiums are usually calculated on yields for a large area such as a shire. Insurers don't know which farmers have higher than average yields and which have lower than average yields. If the insurer pays individual farmers when they achieve less than, say, 65% of average yields for the area, the farmers with less than average yields will often get big payouts and farmers with more than average yields will seldom get payouts. Farmers with lower than average yields will purchase crop insurance and get a bargain. The second type of information is the variability of yields. Farms with less reliable yields should pay higher premiums. However, yields over a wide area are the total production of many farms and are less variable than individual farm yields. Premiums based on area yields will always be too low. With adverse selection, the indemnities paid out by the insurer will exceed the premiums paid in by farmers and high loss ratios will result, as in Canada, the US and other countries. In Western Australia, two consortiums of private companies have offered multi-peril crop insurance (Multi Peril Crop Insurance Task Force 2003). The first scheme had comprehensive cover with most premiums calculated using shire level data. It began in 1974, sold very little insurance, made large indemnity payments on a few policies and ended in 1975. The second scheme had only catastrophic cover with low premiums calculated from an extensive data base of individual farm yields. It began in 1999 after widespread publicity, sold 34 policies, made no indemnity payments and ended in 2000.

Systemic risk is another major reason most crop insurance schemes fail. It is unlikely that any crop insurance program could have survived the recent droughts in Australia. Either governments must underwrite the risks or reinsurance must be bought from global reinsurers to diversify the risks away from agriculture and away from Australia. Given the history of failures, it is unlikely that a multi-peril crop insurance program can be reinsured.

Nevertheless, the Multi Peril Crop Insurance Task Force (2003) investigated the viability of multi-peril crop insurance in Western Australia. First they designed insurance contracts for individual farmers. Except for specialist varieties, virtually all wheat is delivered to receival points managed by one company. This company tracks deliveries back to individual farms and gave the Task Force access to yield data for 9 years on every farm in 8 agro ecological regions around the state. The usable data comprised 1006 wheat farms, about a quarter of those in the state. The Task Force analysed several possible insurance contracts and recommended a contract with a 65% coverage level, a 70% election percentage and a 70% loss ratio. They found premiums ranging from 0% to 14.5% with most farmers paying relatively low premiums, as shown below in Figure 1.

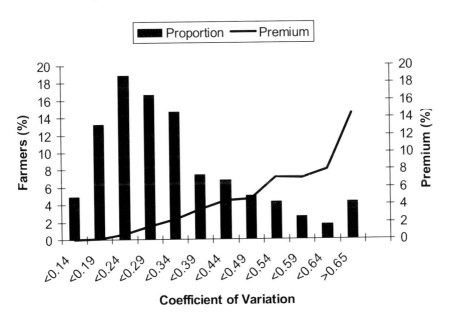

Coefficient of Variation

Fig.1 Proportion of Farms and Insurance Premiums Classified by the Coefficient of Variation (source: Multi Peril Crop Insurance Task Force 2003)

On the left-hand vertical axis is the proportion of farms in different risk categories. On the horizontal axis are risk categories measured by the coefficient of variation, which equals the standard deviation of yields on a farm divided by its average yield. For example, a coefficient of variation of 0.24 says that the standard deviation of yields is 24% of average yields. On the right-hand vertical axis is the premium for multi-peril crop insurance as a percentage of the value of the crop. Most farmers would pay fairly low premiums. This is more easily seen by graphing the percentile of farmers as shown below in Figure 2.

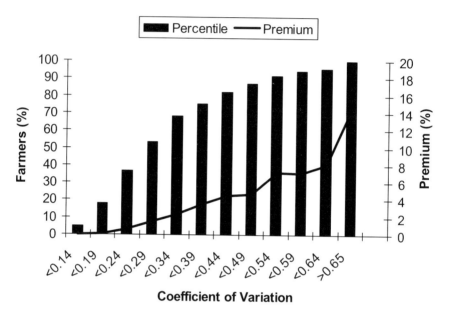

Fig.2 Percentile of Farms and Insurance Premiums Classified by the Coefficient of Variation (adapted from Multi Peril Crop Insurance Task Force 2003)

About 37% of farmers have a coefficient of variation less than 0.24 and would pay less than 0.6% of the value of their crop for insurance. More than 50% of farmers would pay less than 1.5% and 75% of farmers would pay less than 3.5%. These are relatively low premiums, reflecting the reliable wheat production in the state and suggesting that crop insurance would be affordable in Western Australia.

To investigate the potential for adverse selection, the Task Force calculated the premiums that should be paid by the riskiest 10%, 20%, 30% and 40% of farms in each of the 8 shires, as shown in Table 1 below.

There is some variation in premiums among shires. However, there are risky farms in every shire and most of the variation is among farms within shires. An area yield insurance program might set premiums at the average for each shire. The riskiest farmers would consider insurance a bargain, the least risky farmers would consider insurance too expensive and adverse selection would destroy the program. In theory, if not in political reality, insurance could be made compulsory for all farmers. Instead of government subsidies, some farmers could subsidise others. The Task Force calculated a maximum transfer from less to more risky farmers of $14 per hectare per year.

Table1. Premiums for the riskiest farms in each state

		Riskiest Farms			
Shire	Average (%)	10%	20%	30%	40%
Dalwallinu	1.2	5.1	3.8	3.1	2.7
Wongan-Ballidu	1.3	7.6	5.6	4.1	3.1
Dandaragin	1.9	9.2	6.0	4.8	4.1
Katanning	2.2	7.1	5.4	4.0	3.4
Merredin	2.3	7.3	5.1	4.3	4.3
Kulin	2.5	7.9	5.8	5.5	4.9
Esperance	2.6	9.4	7.1	5.8	4.9
Jerramungup	4.1	10.9	9.8	8.3	7.3
Average	2.1	7.7	5.9	4.8	4.2

(source: Multi Peril Crop Insurance Task Force 2003)

Western Australia has the lowest systemic risk in Australia. In the south and west nearer the coast is a high rainfall zone that produces well in dry years and less well in wet years. Far inland in the north and east is a low rainfall zone that produces well in wet years but may produce nothing at all in dry years. In between is an intermediate rainfall zone that produces well in most years. Over the past few decades, Western Australia has had reliable production. Even so, the 1994/95 and 2000/01 crop years were drier and the 2002/03 crop year was very dry with complete crop failures inland. The Task Force investigated the degree of systemic risk in Western Australia by analysing a hypothetical scenario. Suppose a multi-peril crop insurance scheme was established and premiums were set using 5 years of data from 1992/3 to 1997/98. The premiums would have been too low for the scheme to survive the 2000/01 crop year. High premiums could have been charged in early years to weather the poor years, but few farmers would purchase insurance. The only alternative is to reinsure with a financial institution that already has sufficient capital reserves to finance indemnity payments early in the program.

Finally and hypothetically, if the problems of moral hazard, adverse selection, transaction costs and systemic risk could be solved, would farmers purchase multi-peril crop insurance? As yet, there is no definitive answer. The experience in the US and Canada shows that farmers will purchase insurance if the premiums are subsidised, but will farmers pay commercial premiums? Australian farmers have many other risk management tools. Many farmers have Farm Management Deposits as a tax-effective way to save for difficult years. Quite a few farmers have off-farm investments in property and stocks. Some farmers diversify geographically by operating farms in different rainfall zones. Almost all farmers diversify by mixing various crops and livestock on their farms. Yet savings and off-farm investments do not reduce yield risks and will introduce other financial risks. Diversification does not make farming more efficient; crop insurance may. With crop insurance farmers may become more entrepreneurial and adopt new technologies. They may specialise and produce more efficiently. Both the federal Multi Peril Crop Insurance Project and the Western Australia Multi Peril Crop Insurance Task Force surveyed farmers to assess the likely adoption. Both made preliminary assessments that about 18% of farmers would purchase crop insurance at commercial premiums.

4. Rainfall insurance

In Australia, the viability of rainfall insurance was debated almost two decades ago (Bardsley *et al* 1984; Quiggin 1986). Rainfall insurance has several advantages over multi-peril crop insurance. Moral hazard is minimal. Farmers cannot affect the weather, although insurance companies may become insolvent and be unable to pay indemnities. Adverse selection is unlikely. Rainfall data is collected by an independent third party, the Bureau of Meteorology, and is known to both insurers and farmers. Transactions costs are low. Contracts are standardised and assessing crop damage is unnecessary. Systemic risk is easy to manage because the problems of moral hazard, adverse selection and transaction costs are solved, making reinsurance easy to obtain. Finally, although it solves other problems, rainfall insurance introduces basis risk. A farm's yield is imperfectly correlated with rainfall. In some years a farm may have acceptable yields and still receive an insurance payout. In other years a farm may have poor yields and not receive a payout.

Basis risk can be explained with the help of Figure 3 below.

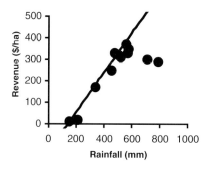 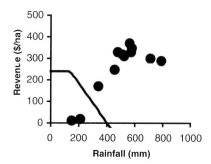

(a) Predicted Revenue   (b) Linear Payouts

Fig.3 Basis Risk from Rainfall Insurance

In panel (a), actual and predicted revenues, in dollars per hectare ($/ha), are plotted versus rainfall, in millimetres (mm). Actual revenue is shown by the dots. Predicted revenue is shown by the straight line and equals the predicted rainfall multiplied by a predicted crop price. Rainfall insurance is a contract written using predicted instead of actual revenues. Suppose average rainfall is 500 mm and the coverage level is 80% of average, or 400 mm. At 400 mm, revenue is predicted to be $240/ha. At 125 mm, revenue is predicted to fall to $0/ha. In panel (b), the payout is calculated as $240/ha minus the predicted revenue. The payout is $0/ha at 400 mm and rises to $240/ha as rainfall falls to 125 mm. For lower rainfall, payouts are capped at $240/ha. Because rainfall insurance approximates actual revenue, insurance payouts are imperfectly related to actual damages and there is basis risk.

Although many variations on multi-peril crop insurance have been implemented around the world, rainfall insurance is less common. Currently in eastern Australia, insurance adjustors are willing to sell rainfall insurance that is backed by financial institutions. The details are commercial and in-confidence, and few policies have been sold (personal communication). Although rainfall insurance is ideal for insurers because it solves the problems of moral hazard, adverse selection and transaction costs, and makes systemic risk easier to manage, it is less useful to farmers because of basis risk.

5. Weather derivatives and yield index insurance

Rainfall insurance has recently been reborn as weather derivatives. Weather derivatives are sold by financial institutions and purchased by municipalities, energy companies and tourist industries as a hedge against inclement weather. 'Derivative' describes a financial product that is derived from something, almost anything, else. For example, the price of a futures contract for wheat is derived from the price of wheat. Even further, the price of an option on futures is derived from the futures price that was derived from the price of wheat. Weather derivatives are derived from millimetres of rainfall or degrees of temperature at Bureau of Meteorology weather stations around Australia. Long and reliable data series allow premiums to be calculated with confidence. Compared to rainfall insurance, transaction costs are lower and systemic risk is reduced because financial institutions require fewer capital reserves and are more widely diversified than insurance companies. Like rainfall insurance, however, weather derivatives are an approximation and have basis risk.

Yield index insurance (Quiggin 1994; Skees 1999) is a way to reduce the basis risk, as shown in Figure 4 below.

A yield index is a non-linear model of yields as a function of rainfall. The better the prediction of actual yields, the lower the basis risk. In panel (a), the yield index is multiplied by a contract price and converted to revenue. Yield index insurance is a contract written on the predicted revenue. Instead of payments triggered by low rainfall, payments are triggered by low revenue. Suppose the coverage level is $240/ha. If rainfall is 500 mm, revenue is predicted to be $315/ha. The farmer gets no payout, regardless of actual revenue. If rainfall is 400 mm, revenue is predicted to be $225/ha and the farmer receives a payout of $15/ha. In panel (b), payouts begin at 415 mm of rainfall and rise to $240/ha as rainfall falls to about 100 mm. At the other extreme, too much rainfall is damaging as well, and the farmer begins receiving payouts as rainfall rises above 840 mm.

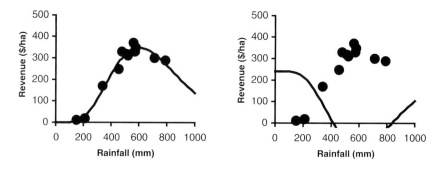

(a) Predicted Revenue   (b) Non-linear Payouts

Fig. 4 Reducing Basis Risk with Yield Index Insurance

Weather derivatives are ideal for financial institutions but may have little relevance to farmers if basis risk is too large. Rainfall at a Bureau of Meteorology weather station may be a poor predictor of yields on a farm. Yield index insurance reduces the basis risk. Unfortunately, it is more complex and, in financial jargon, is an 'exotic option' (Zhang 1998), so exotic that we know little about it. Yet yield index insurance may be more relevant to farmers, other rural businesses or even rural towns.

6. Prospects for the future

Multi-peril crop insurance has not solved the problems of moral hazard, adverse selection, transaction costs and systemic risk and is not commercially viable. Australian governments are unwilling to subsidise crop insurance premiums or underwrite yield risks. Hence, the only prospects for insuring crop yields in Australia are weather derivatives and yield index insurance. Financial institutions are selling weather derivatives in the Northern Hemisphere and would like to diversify to the Southern Hemisphere (AXA Australia, personal communication).   Australia has reliable weather data and a large research program on managing climate variability (White 2000a). Hence, transaction costs will be relatively low. To lower costs further, financial institutions would prefer multi-million dollar contracts. The challenge is to bridge the gap between yields on the farms of Australia and multi-million dollar contracts with global financial institutions.

Building the bridge requires knowledge about:

- How to estimate simple but accurate yield indexes for individual farms;
- How to set the premiums for yield index insurance;

- How to determine the demand by farmers for yield index insurance in a portfolio of Farm Management Deposits, off-farm investments and diversified production;
- How to pool yield index contracts over several farms for reinsuring with financial institutions.

Since there is no problem with moral hazard or adverse selection, farmers' own yield histories can be combined with rainfall and temperature data from nearby weather stations to estimate yield indexes. Many complications may arise, however. Rainfall and temperature data at weather stations may be poorly correlated with those on farms. Two or three weather stations combined might give a better correlation but this will require close examination of the data. Yields are sensitive to rains at planting and filling of the grain, but less sensitive to total rainfall for the growing season. Both heat and cold interact with rainfall. The weather data that best predicts yields may be quite complex. Further, robust statistical methods must be applied to estimate robust yield indexes, and it is a job for statisticians. Yet the resulting indexes must be understood by farmers, farm advisors and insurers and be simple to use.

Robust methods for setting insurance premiums and for pricing financial derivatives are based on the probabilities of different outcomes. In addition, methods for pricing financial derivatives are based on the assumption that the derivatives will be freely traded in a market. Weather derivatives and yield index insurance, however, will be designed specifically for each farm and will not be traded. Hence, they will be less flexible and must be priced accordingly. Yield index insurance is written on a non-linear prediction, rather than directly on weather itself, and is even more difficult to price correctly. Robust methods for pricing weather derivatives and yield index insurance are yet to be developed.

The likely demand for insurance can be assessed by surveying farmers. This seems easy, but it will be difficult in practice. A yield index and a price must be calculated for each farmer in the survey using the farmer's own data and the nearby weather stations. Given this information, a farmer's demand for insurance will depend on all other decisions in their portfolios, on their wealth and ability to bear risk and on their attitude toward risk. A farmer who is less able or unwilling to bear risk will buy more insurance. Farmers in Australia have no experience with crop insurance and will need help in learning about it and in deciding whether and how it fits in their portfolios. For help they may turn to their farm advisors. Therefore a survey of farmers will be an intensive effort by researchers, farmers, farm advisors and facilitators.

If likely demand is sufficient, an intermediary between farmers and financial institutions must be created. The intermediary could be a government. For example, in Canada, the provinces run the crop insurance programs, in cooperation with the federal government, and are responsible for reinsuring the risk. The province of Alberta constructs a vegetation index using satellite imaging and weather station data and pays indemnities to cattle ranchers based on this index. Alberta also constructs a rainfall index from 17 weather stations and buys weather derivatives from a financial institution as reinsurance (Multi Peril Crop Insurance Task Force 2003). Alternatively, the intermediary could be a bank, a stock firm or a grain trading company. Finally, the intermediary could be a new generation cooperative or mutual fund that is owned and operated by farmers or rural businesses.

In conclusion, the welfare of society will improve if farmers can specialise and produce according to their comparative advantage. Yield risk forces farmers to diversify and produce less efficiently. If commercially viable crop insurance can be invented, farmers will be able to hedge their risks and free themselves to produce efficiently. The best prospect is yield index insurance derived from weather data, a pool of yield index contracts managed by an intermediary and reinsurance for the pooled portfolio from a global financial institution. Nor should it stop there. Insurance may also be demanded by intensive piggeries, machinery dealers, stock firms, banks or even by rural towns.

Drought policy in Australia includes Exceptional Circumstances, Farm Management Deposits and climate research through Land and Water Australia, the Bureau of Meteorology and state departments of agriculture. It could also include the creation of yield index insurance, building upon Australia's expertise with rainfall insurance and climate risk management.

CHAPTER 10: POLICY FOR AGRICULTURAL DROUGHT IN AUSTRALIA: AN ECONOMICS PERSPECTIVE

BRUCE O'MEAGHER*

Consultant, 169 Wattle Street, O'Connor, ACT 2602, Australia

1. Introduction

This chapter examines agricultural drought from an economics perspective. It summarises the economic implications of drought, its economic impacts and related costs. The appropriate roles of the private and public sectors in relation to drought policy are discussed and the relative efficiency and effectiveness of government intervention under Australia's National Drought Policy (NDP) is assessed. It is argued that governments have an important role to play but that the primary responsibility for drought policy rests with private individuals and enterprises. It is concluded that current government policy intervention is ineffective and undermines economic efficiency and equity. Some suggestions are made for policy improvement.

2. Economic aspects of agricultural drought

Drought is a normal feature of climate variability experienced to some extent by all countries. Although there is no universally accepted definition, there is common agreement that it originates from a deficiency of precipitation over an extended period, the duration of which differs from country to country and often from location to location within countries (Wilhite 2000a). There is common agreement also that, with the possible exception of the impacts of our contribution to greenhouse gas emissions on the intensity of climate variability, drought is a natural phenomenon over which we humans have little influence and no control. As such, it represents a significant constraint on a broad range of human activities, including agricultural activity (Wilhite and Vanyarkho 2000). The literature distinguishes between meteorological, agricultural, hydrological and socioeconomic drought (Wilhite 2000a).

Unlike our reaction to other natural hazards, our realisation that drought onset has occurred emerges only slowly and, when that realisation occurs, we have little real idea of how long it will last and what its longer-term impacts are likely to be. Agricultural drought may therefore be said to involve considerable uncertainty and risk. Hardaker, Huirne and Anderson provide a useful distinction between these terms, defining '*uncertainty* as imperfect knowledge and *risk* as uncertain consequences, particularly exposure to unfavourable circumstances' (Hardaker *et al* 1997, p7).

*The author wishes to thank John Freebairn and David White for their helpful comments on earlier drafts of this chapter. Don Brunker helped to clarify a number of points of interpretation. The views expressed are those of the author and may not necessarily be shared by referees.

L.C. Botterill and D.A. Wilhite (eds.), From Disaster Response to Risk Management, 139–155.
© 2005 *Springer. Printed in the Netherlands.*

Our uncertainty about drought and its associated risks stems from several sources, including imperfect knowledge of climate dynamics and hence drought incidence; interactions between meteorological, hydrological and agronomic systems and hence of the precise impacts individual events will have on farm production and income; and likely market-based responses to drought onset, including by other producers (at home and abroad), input suppliers and financial institutions.

There is a considerable literature on Australia's experience of drought, with valuable overviews provided by Foley (1957), Gibbs and Maher (1957), Lovett (1973), Chapman (1976), (Heathcote 2002) and Botterill and Fisher (2003). Drawing particularly on Stafford Smith (2003b) and Lindesay (this volume), key factors in the Australian context include:

- Australia is located in a particularly variable part of the earth's climate system and is subjected to the influence of several climate sub-systems;
- under the influence of these and other factors, Australia experiences considerable seasonal variation in rainfall across its several climatic zones; moreover, for some regions the pattern of seasonality has changed over the past century or so;
- Australia has the highest rainfall variability and run-off of any continent, with variability generally increasing inland and those areas experiencing the highest variability tending to be those with the lowest mean rainfall;
- as a result of climate variability and generally poor soils, agricultural activity is predominantly extensive and the general environmental stock relatively vulnerable; and
- drought incidence and intensity is rarely uniform, including during spatially extensive drought episodes.

2.1 ECONOMIC CHARACTERISTICS OF AGRICULTURAL DROUGHT

From an economic perspective, agricultural drought may be regarded as an inevitable exogenous (or external) variable which, through its impacts on hydrological and agronomic systems, can impose significant constraints on production and income possibilities of drought-affected agricultural enterprises. Though a source of considerable risk for these enterprises, it is only one of an array of risks faced, all of which need to be factored into enterprise business risk management planning.

Drought impacts vary according to location and enterprise type and the effectiveness of the risk management practices adopted. Droughts therefore create winners as well as losers—but this drought's winners may be the next drought's big losers. Drought risk management expenditures by farmers represent incomes to upstream and downstream businesses within the agriculture sector, for example for business and property planning advisers, and to the regional communities in which their farms are located. Even after drought onset, relevant services continue to generate income within regional communities. Importantly, drought-induced price increases for farm outputs may benefit those farmers unaffected by drought whereas increases in input prices may hurt them. Perversely, given the fickle nature of rainfall variability, sometimes there are lucky farmers within a predominantly drought-affected region who receive timely, adequate rainfall.

Enterprise impacts (both adverse and positive) flow through to the communities in which they are located, into regional economies and, for significant events, potentially the national economy. This inevitably involves enterprise, intra- and often inter-sectoral adjustments. There can also be global implications through internationally traded volumes and prices. Here too there can be winners and losers—drought in Australia may result in a fall in export earnings one year followed by increased earnings derived from another country's drought the next.

A point worth noting in the global context is that the scale of drought impact is likely to be very different in developed countries compared with the impacts in lesser developed countries, particularly those which are heavily dependent on agriculture as a source not only of food but also of national income. Unlike developed countries with diversified and well-developed markets and welfare systems able to absorb the adjustments and income losses involved, lesser developed countries can find themselves extremely exposed to even mild drought events (Benson and Clay 2000).

2.2 ECONOMIC IMPACTS AND RELATED COSTS

Discussions of enterprise and sectoral impacts are included in several of the overview references referred to earlier. Stafford Smith *et al* (1997; 1994b) and Hammer *et al* (2000) provide further references while economic overviews can be found in Anderson (1979) and the annual farm survey reports published by the Australian Bureau of Agricultural and Resource Economics (ABARE) and its predecessors and by the various state departments of agriculture.

Given patterns of normal seasonal variation noted earlier (see Lindesay, this volume), short, sharp drought events may have little impact on some enterprises. But they can have significant impacts on both broadacre cropping and livestock enterprises if they 'hit' at just the wrong time—at critical seasonal junctures such as maturation for wheat crops or during lambing for sheep operations. Droughts of longer duration can be expected to hit hard in the early stages for cropping operations (yield reduction or crop failure) compared to grazing operations which may be hit harder during the latter phases of a drought episode and, particularly where stock numbers (including breeding stock) have been reduced, during a much longer period after the drought has broken.

Farm sector impacts for major events can be significant, although their longer-term impact on farm business performance will be strongly influenced by a combination of how well farm businesses are positioned before drought onset and how well they are managed after onset. Purtill *et al* (1983, p5) report that more than 60% of agricultural and grazing operations were drought declared during the 1982-83 drought, although there was considerable geographic variation, with Western Australia virtually unaffected. The volume of rural output was estimated nationally to have declined by 18% during the year largely as a result of the drought (Campbell *et al* 1983, p255) while farm receipts were estimated to have declined by around 23% and farm cash operating surplus by around 50% (Campbell *et al* 1983, p254; Purtill *et al* 1983, p11).

Similarly, around 50% of broadacre farms (mainly in the eastern states) were drought affected during 1992-94, some in both years (Martin 1995, p60). However, there were significantly different levels of impact within eastern Australia between the two droughts, with the major impacts in 1982-83 falling on the southeast while the major impacts in 1992-94 fell in northern Australia. South Australian and Victorian farms were both less affected by the 1992-94 drought. Martin notes (1995, p60), however, that estimated aggregate real farm business profit averaged -$38,000 in the 1982-83 drought compared to -$13,700 in 1994-95. A significant difference between the two droughts was the fact that farm enterprises entered the 1982-83 drought in good shape whereas farm businesses entered the 1992-94 drought in poor shape, largely as a result of poor commodity prices in the preceding period. On average, for the 1992-94 drought, Martin notes (1995, p61) that incomes for farms affected by drought were estimated at about 50% below those which were not affected while those farms which had been affected in both years were estimated to have an income only around 17% of those unaffected.

Experience in the recent 2002-03 drought, which was more extensive than the 1982-83 and 1992-94 droughts, illustrates the differential impacts which drought can have on different farm enterprises (Martin *et al* 2003, pp1-14). Production of winter grain crops in 2002-03 was estimated to be down by 60% to the lowest level since 1994-95 while summer crops were down by 50%. Overall crop receipts on broadacre farms were estimated to have fallen by 50%. Livestock production was also adversely affected, with the potential positive income effects of increased turnoff being offset by lower saleyard prices resulting in receipts for livestock broadacre enterprises estimated to have fallen by some 15%. In contrast, wool receipts were estimated to have increased by around 16% in 2002-03 as a result of higher wool and sheep meat prices having offset the lower wool production resulting from the fall in sheep numbers and wool cut per head.

Overall, broadacre farm cash incomes during 2002-03 are estimated to have declined by around 56% to similar levels experienced in the early 1990s while farm business profits fell from $42,720 to -$46,300. The flow through impacts on farm enterprises, however, could be expected to be cushioned to some extent by the fact that farms entered the drought with relatively high equity levels as a result of two very good income years, with 2001-02 incomes being the highest on record (Martin *et al* 2003, p4).

These enterprise impacts can have significant macroeconomic implications. Reductions in GDP as a result of drought have been estimated at 1.1%, 0.75% and 1% for 1982-83 (Campbell *et al* 1983, p257), 1994-95 (Munro and Lembit 1997, p6) and 2002-03 (ABARE 2003). Other variables can also be significantly affected, including employment (estimated 2% reduction in 1982-83), trade (estimated 2% reduction in 1982-83) and inflation (estimated increase of 0.2 percentage points in 1994-95).

As will be discussed in the following section, drought response strategies represent significant costs to government as well as to farm business enterprises. Australia's national government is estimated to have provided almost $500 million in support to drought-affected farmers between 1993 and 1996 while state governments provided a further $190 million (Munro and Lembit 1997, p10). National government drought support expenditure for the 2002-03 drought is estimated at $1.2 billion (Treasury 2003, pp73-4).

3. Public policy responses in Australia

Although we are unable to control the frequency, duration and intensity of drought, its impacts can be influenced through the decisions and actions we take. We are principally concerned to reduce uncertainty (mainly through research and information and technology transfer) and risk (through the various response strategies we adopt). These responses can have positive and negative outcomes, depending on the choices made. How best to respond to drought therefore raises important policy considerations for individuals, enterprises and governments alike, particularly for countries like Australia which experience significant climate variability.

As we have just seen, the scale of public response is significant, though it has not always been so. Overviews of the development of drought policy in Australia are provided by Drought Policy Review Task Force (1990), O'Meagher et al (2000) and Botterill (2003d and this volume). These developments are usefully seen within the context of broader agricultural policy development, particularly post-war development.

As for other countries, income instability associated with the array of risk factors facing agriculture has been a major preoccupation with Australian agricultural policy. This has involved significant government intervention by national and state governments, principally justified on the basis that such intervention would contribute to increased price and income stability, economic efficiency and the welfare of rural and regional communities. Principal intervention measures included a range of domestic subsidies and tariffs and import quotas, all of which had the effect of distorting production and price signals. Importantly, similar measures were applied to manufacturing (Butlin *et al* 1982).

A range of factors contributed to the questioning of across the board protection, particularly during the 1970s and 1980s, principally the realisation that such strategies were unsustainable in a rapidly changing international economic context (Anderson and Garnaut 1987). A process of substantial microeconomic reform was begun fitfully during the 1960s and 1970s and gathered pace during the 1980s and nineties (Industry Commission 1998). Although there are differing views (see, for example, Pusey 1991; Pusey 2003), there is substantial evidence that the process of broad microeconomic reform has delivered substantial benefits in Australia's overall economic performance and well-being (Gruen and Shretha 2000; Parham 2002).

Agricultural reform has been a significant component of that process of microeconomic reform. Assistance to agriculture has been reduced significantly (Industry Commission 1995; Industry Commission 1998; Productivity Commission 1998) and Australia now has the second least protected agricultural sector within the OECD (Organisation for Economic Cooperation and Development 2003).

Drought policy was not immediately caught up in this reform process in part because of prevailing perceptions about drought as a natural disaster rather than a risk management factor (Heathcote 1973; O'Meagher et al 2000) and partly because of concerns about farmer welfare. Government support was principally event specific, although drought research was being undertaken by the CSIRO and the Bureau of Meteorology, taxation concessions were extended for water conservation measures, and drought management did feature to some extent in ongoing agricultural extension services. The principal forms of short-term, event-specific assistance comprised interest rate, agistment, transport, fodder and water cartage subsidies.

These interventions were generally justified on the basis that they addressed what were claimed to be the special circumstances of agriculture—fluctuating incomes, relative geographic isolation, essential industries etc. Economic arguments included that such intervention contributed to efficiency by: helping to reduce the costs of retaining resources during times of income instability (thereby contributing to longer-term agricultural sector stability and investment); offsetting the costs imposed by support extended to agricultural inputs (tariff compensation); addressing weaknesses in financial markets; and facilitating smoother structural adjustment (Freebairn 1978).

The mounting cost, evidence of 'rorting', accelerated scientific understanding of the factors driving climate variability and a determination by some key decision makers to ensure that drought policy did not undermine broader reform strategies eventually led to the withdrawal of drought from the natural disaster relief arrangements under which drought support was provided, a national enquiry into appropriate drought management policy approaches and the adoption of the National Drought Policy (NDP) in 1992.

3.1 THE NATIONAL DROUGHT POLICY

The detailed evolution of Australia's NDP has been discussed extensively elsewhere (Botterill 2003d and this volume; White et al, this volume; Burdon 1995; O'Meagher 2003; O'Meagher et al 1998; O'Meagher et al 2000; White and Bordas 1996; White and Karssies 1997).

In broad terms, the 1992 announcement represented a significant rhetorical departure from the past. In summary, event-related support was only to be provided in exceptional circumstances; public intervention was to emphasise risk management, preparedness and the self-reliance of farm business enterprises rather than coming to the aid of 'farmers in trouble'. Importantly, fodder and other subsidies (but not interest rate subsidies) were to be phased out. Other policy elements included an information dissemination program, increased support for education and training activities and providing more generous arrangements under the Income Equalisation Deposits scheme, the national government's savings incentive scheme for agricultural enterprises (later re-characterised as Farm Management Deposits or FMDs). Soon after the NDP was implemented, the policy was significantly extended by granting farmers and their families experiencing exceptional drought conditions access to generous welfare assistance previously denied under the general welfare arrangements. Broader agricultural support measures (such as tax averaging) were to be retained.

Despite the new political rhetoric of self-reliance and risk management, however, there has been significant continuity in actual implementation policies. And as the above references have pointed out and agriculture Ministers themselves have acknowledged (ARMCANZ 1997), as the policy has continued to evolve, the intention of establishing a national focus and the ideal of self-reliance has proved remarkably elusive. Key NDP developments have included:

- transaction-based subsidies and concessional loans have been retained by some states;
- interest rate subsidies provided for farm business support have been extended to small businesses in drought affected areas;
- access to the welfare system introduced in 1994 has resulted arguably (see below) in a shift in emphasis from input subsidies to subsidies in the form of welfare payments;
- administration of the exceptional circumstances declaration procedures has been softened (including the introduction of interim assistance while applications for exceptional circumstances are being assessed) and, as a result, has undermined the welfare-specific intent of welfare payments (O'Meagher *et al* 1998; White *et al*, this volume);
- broad objectives of the policy have been diluted by successive revision and Ministerial statements (Botterill 2003d).

3.2 OBJECTIVES AND MEASURES

Against this background, the accompanying box provides a summary of the main drought policy objectives identified by national and state Ministers since 1992 and the main intervention measures which have been adopted to achieve those objectives. These provide the basis for the economic assessment of current drought intervention in the following section.

4. Economic assessment

4.1 CONTEXT

From an economics perspective, assessment of public policy intervention for drought should be based on the same criteria as for any other area of policy intervention— namely, whether there is a clear and demonstrable market failure justifying intervention and whether specific intervention strategies are likely to result in overall net gains to the community through improvements to efficiency in resource allocation and/or to welfare improvements, especially improvements in social equity outcomes (Kraft and Piggott 1989; O'Meagher 2003). Arguments associated with broad economic stability are now generally accepted as falling within the ambit of macroeconomic management, with which we are only peripherally concerned here.

AUSTRALIA'S EVOLVING NATIONAL DROUGHT POLICY *

Objectives
- Move from 'band aid' support to drought risk preparedness
- Increase farm business enterprise self-reliance based on the adoption of integrated (including drought) risk management
- Promote sustainable natural resource use
- Separate farm business support and welfare measures
- Provide income support to those farmers unable to meet living expenses, but without undermining economic adjustment
- Provide a basis for early drought recovery

Primary Measures
- Direct expenditures
- Research
- Information provision
- Welfare support
- Subsidies
    - Interest rate subsidies
    - Transaction-based subsidies (especially transport/fodder subsidies)
    - Farm Management Deposits (FMDs)
    - Education and training subsidies

* This table draws in part on Simmons 1993.

The presumption of market failure as a basis for public intervention is sometimes based on pure welfare economics propositions about the benefits to be derived from perfectly competitive markets; more often (as here) it is based on the pragmatic observation that government failure is extensive and that reasonably competitive private markets can be relied upon to deliver workably efficient 'second best' outcomes at less cost to the community (Kay 2003; Lindblom 2002). Importantly, government plays a critical role in providing the legal, institutional and regulatory environments which are necessary to enable workably efficient private markets to operate.

Private markets may, for a broad range of reasons, fail to deliver the goods and services the community desires or, alternatively, may be unable to deliver these goods and services at a lesser cost than the public sector. Reasons for market failure include the lack of sufficient incentives to deliver public goods (for example, welfare services), the presence of structural impediments to competition (for example, monopoly), the presence of externalities (for example, adverse environmental consequences from private, market-based activity) and information failures (for example, lack of relevant information to enable effective decision making).

Although private markets can fail, government intervention may also fail, not least because they more often than not face the same kinds of constraints (such as information constraints) that cause private market failure. Other sources of failure include political and bureaucratic 'inefficiencies' (for example, personal ambitions) which bear little or no relation to overall community welfare. Although not of prime concern here, it may be noted that government failure has been a feature of the history of drought intervention over many years in Australia (Heathcote 2000; O'Meagher 2003; O'Meagher *et al* 2000).

Against this background, market failure is usually regarded as a necessary but not sufficient reason for intervention since intervention should also result in net benefits to the community as a whole in the form of efficiency and/or equity gains. Importantly, efficiency gains are taken to mean gains (or at least no efficiency loss) to the whole community, not just to a specific industry or sub-sector of industry. It is little use in improving efficiency in resource allocation in one sector if, as a result, it has net negative allocation implications for society as a whole.

Though principally concerned with efficiency considerations, economists are also concerned with equity considerations since some interventions for equity reasons may be more efficient than others and because equity-based intervention may have negative impacts on efficiency. For this latter reason, there are frequently trade-offs which need to be considered between efficiency and equity goals.

4.2 SHORT-TERM SECTORAL ASSISTANCE

A striking feature of current public policy for drought is its overwhelming short-term, event-oriented focus. Although there are aspects (such as research and extension) which are of a longer-term nature, expenditures and political and bureaucratic attention are dominated by interest rate and transaction-based subsidies and temporary welfare services and payments.

Freebairn (1978; 1983) and others have discussed the absence of market failures justifying subsidy support and why such support is unlikely to contribute to improved efficiency and equity. McColl *et al* (1997) notes that a national finance systems inquiry found no evidence of market failure in financial services to the rural sector and that a Reserve Bank survey had reported that rural borrowers were generally more satisfied with the services provided by the banking system than were metropolitan borrowers. And as the Industry Commission (Industry Commission 1996, p30) has pointed out, the fact that some farmers experience difficulties in accessing finance is not in itself an indication of market failure but rather a reflection of the commercial judgements made by lenders about the credit worthiness of potential borrowers. Nor is there any substantive evidence of market failure in respect of fodder trade—to the contrary, the evidence of recent significant drought events points to an extensive and responsive market. Fodder prices do rise during drought episodes, but this reflects overall demand and supply situations resulting from drought extent and duration and farmers' individual decisions (including decisions regarding the creation of buffer stocks) rather than market failures.

Traditional efficiency-based defences for such interventions have long been questioned by economists and are now largely (though not entirely) abandoned even by rural interest groups. Freebairn (1978) has demonstrated that there are unlikely to be significant under-investment or resource flow costs arising from the adverse impacts of climate variability on longer-term expectations concerning enterprise returns. Fundamentally, such variability is a fact of life which is factored into longer-term expectations and contingency planning. There are few if any grounds for expecting that governments are more likely to be in a better position to do this than individual farmers. Where farm adjustment occurs as a result of ineffective private contingency planning or decision making, there are unlikely to be significant net efficiency costs to the community since sector specific assets are likely to be retained for use within the sector while mobile factors (including labour) are likely to be readily able to be employed by other industries. Nor are such subsidies likely to be effective tools for smoothing the adjustment process since they distort market prices and can create expectations that government will come to the rescue of 'farmers in trouble' during the hard times. Such expectations have the effect of not only undermining the focus on the need for effective risk management planning and self-reliance but also impeding adjustment itself by encouraging resource 'stickiness' and the preservation of outmoded industry structures (Freebairn 1978; O'Meagher 2003). Both forms of subsidy may also increase natural resource vulnerability by providing incentives to overuse fertilisers or retain stock for longer than would be the case under integrated risk management planning (Freebairn 1983).

In the absence of evidence of market failure and the likely adverse impacts on efficiency, there remains consideration of whether such short-term intervention is likely to improve equity or result in improved welfare. Our discussion is confined to interest and transaction-based subsidies for the moment with consideration of general welfare system payments discussed in a later section.

Drawing in particular on Freebairn's (1983) examination of the impacts of interest and fodder subsidies during the 1982-83 drought, a number of objections may be raised to these kinds of subsidies on equity grounds. There is no doubt that, notwithstanding the potential for efficiency-based externalities referred to above, such subsidies may assist the individual farmers who are able to access them. On the other hand, they are only available to drought-affected farmers; they are unlikely to be accessed by those farmers who have more successfully planned for variability, including by building their own financial and fodder reserves; although they may have positive impacts on financial institutions and some fodder producers and transporters, they may have adverse impacts on the prices other drought- and nondrought-affected farmers have to pay for finance and fodder. Other sectors of the economy which are affected by climate variability (such as tourism, construction or other primary sectors) are generally unable to access similar support; and they impose efficiency costs on the rest of the community, including potentially through adverse commodity price movements.

Against this background, the use of interest and transactions-based subsidies must be considered as very blunt instruments for improving the welfare of drought-affected farmers. If such concerns are justified, they are more appropriately addressed by direct welfare payments. At the end of the day, the current short-term subsidies paid for drought relief are unlikely to result in overall efficiency or equity improvements and are not justified. In the absence of clear and demonstrable market failures, they have the result in the longer term of increasing the returns to agriculture above those that would have resulted from normal market operations. In this context, Freebairn has commented:

> Such subsidisation causes over the longer run too much labour, capital and other scarce resources to be drawn into agriculture away from other parts of the economy. Australia's world class economic growth over recent decades has been achieved in part by removing selective industry assistance. Drought assistance would be a retrograde move for a productive economy. (Freebairn 2002b)

4.3 RESEARCH AND EXTENSION

Scientific, economic and social research is essential to the task of reducing the uncertainty and risks associated with drought. The knowledge creation that may be derived from research generally is broadly recognised to have significant implications for innovation and economic growth (Industry Commission 1995; Organisation for Economic Cooperation and Development 2003). Moreover, there is well-recognised potential for market failure associated with research activities, particularly involving public good and significant spillovers (Industry Commission 1995). Nevertheless, as the Industry Commission points out, such failures only justify intervention where intervention is likely to result in sufficiently high social payoffs and the research may not otherwise take place.

Although it will be recognised that there are strong public good characteristics of much drought research, it should also be recognised that this is not so in all cases. For example, market failure is less likely in areas of applied research with strong potential for commercialisation. But even here there can be a degree of ambiguity. Research into drought-resistant trees, pastures and crops where the property rights over genetically engineered species are vested in the hands of a private company rather than in the community is an example of where public intervention may not be justified. On the other hand, there may be strong arguments favouring intervention in such research programs to ensure public safety. Where ambiguities of this kind arise they are likely to be resolved through the political process, although economic analysis can contribute to evaluating the costs and benefits of different courses of action. They may also result in public-private research partnerships where the resulting intellectual property rewards are appropriately shared.

Drought-related research in Australia is extensive, largely publicly funded, and being undertaken by a broad range of (mainly) public and private institutions and individuals (Beynon *et al* 2000; White *et al* 1999a). Public sector involvement includes the research sponsored by the national government's research and development corporations and undertaken by other research organisations such as the Bureau of

Meteorology, the CSIRO and the Bureau of Rural Sciences, state departments of agriculture and publicly funded universities.

It is difficult to assess with any precision the efficiency gains which have been derived from this research. Although it is noted that it has provided the basis of our improved understanding of climate variability and its associated risks and has yielded many of the tools which we now use as a basis of drought risk management, no empirical evidence is available in terms of benefit/cost analysis and broader community returns. Issues concerning how much and what kind of research is to be undertaken are still largely resolved outside the marketplace—that is, within the political and bureaucratic process. Whether this is delivering optimal outcomes is largely unanswerable even though there can be little doubt that it is very likely to have resulted in net overall benefits.

Similar considerations, in general, apply to agricultural extension—defined by Marsh and Pannell (2000, p607) 'to include public and private sector activities relating to technology transfer, education, attitude change, human resource development, and the dissemination and collection of information.' Without such activity, it is highly unlikely that farmers and the broader community would be in a position to benefit from the knowledge creation derived from research.

There is broad agreement that market failures can characterise many aspects of extension services (Marsh and Pannell 2000; Mullen *et al* 2000). Accordingly, much of Australia's drought-related extension is provided by public sector organisations. However, several commentators have pointed to the growth of private sector provision in the area of agricultural extension generally, particularly as governments have changed the balance between their research and extension activities (DPIE 1996; Marsh and Pannell 2000; McColl *et al* 1997; Mullen *et al* 2000). Marsh and Pannell have argued that this points to clear evidence of crowding out of private sector intermediation between researchers and farmers. They have also noted that one of the difficulties in applying market failure criteria to this area is that all extension relates in one way or another to information 'which always has public good characteristics to some degree, and can always be claimed to be reducing uncertainty, ignorance and misinformation.' They go on to point out that 'Applying the criterion then comes down to assessing degrees of market failure, which is not often easy to judge' (Marsh and Pannell 2000, p615). Cost recovery for publicly provided services may therefore, in appropriate circumstances, provide a basis for contestability and determination of whether services could more effectively be delivered by the private sector. Whether such services are provided at least cost is complicated, however, by the fact that much extension activity is a joint product of research.

The case for public involvement in extension may therefore be somewhat more contentious than is the case with research. The expansion of research data availability and improved understanding of alternative technologies together with lower cost information and communication technologies provide the basis not only for private sector activity in this area but also for farmer uptake. Innovation may thus be changing the extent to which market failures characterise extension (and possibly research) activities.

4.4 FARM MANAGEMENT DEPOSITS

Farm Management Deposits (FMDs) are tax advantaged, interest-bearing savings which may be drawn down in years of low income (including as a result of drought). They may be regarded as subsidies on savings, the objective of which is to help smooth income fluctuations. They are more flexible than similar products offered by the national government in the past. As a result, they are more popular. In the three-year period to end 2002, the number of participants in the scheme rose from around 7,500 to 39,537 (14% of those eligible) and the value of deposits rose from $250 million to $2 billion (Australian National Audit Office 2003; Douglas *et al* 2002). These were generally very good income years. Uptake rises significantly with income; they are used mainly for tax and income smoothing while the main factor influencing farmers not to make deposits is the lack of sufficient income (Douglas *et al* 2002). The loss to revenue from the associated tax expenditures is estimated in 2002-03 at $410 million and, reflecting the impact of withdrawals as a result of the 2002-03 drought, $180 million in 2003-04 (Treasury 2004).

It will be immediately apparent that FMDs are just another subsidy, in this case for savings. Though they are provided on an ongoing basis, they are intended to deal with event specific situations. They are therefore subject to similar qualification on efficiency and equity grounds as other short-term, event-oriented measures. There is no evidence of failures in the savings market and although they are available to most farmers (subject to agricultural income thresholds), they are not available to other sectors subject to income variability including as a result of drought. At the same time, the evidence adduced by Douglas *et al* suggests that they are highly regressive and do not appear to address the needs of those least able to manage income stability.

A justification sometimes given for the retention of FMDs but yet considered is that they compensate for market failure in formal insurance markets. Evidence concerning the latter has been extensively reviewed (Hardaker *et al* 1997). Broadly, there is agreement that multi-peril agricultural insurance is unlikely to be successful without significant government support given the presence of significant moral hazard and adverse selection issues. Such support has proved costly and fraught with potential for significant government failure. There are instances of specific event insurance successfully provided by the private sector (Gudger 1991) but neither multi-peril nor drought-specific insurance has been offered in Australia. Most reviews of the Australian situation point to the problem of moral hazard arising in part from a long history of drought-related support undermining both potential demand and supply of insurance-based products for drought (Mayers 1995). Leaving the moral hazards arising from government drought support to one side, past constraints may have been attenuated by improvements in our understanding of climate dynamics, improved data and increases in the sophistication of insurance products. Mayers (1995; 1996) and Hertzler (this volume) are positive about the potential for private sector innovation and the prospects for provision of event-specific insurance.

Notwithstanding the incomplete nature of formal insurance markets for agriculture generally and drought in particular, this may not of itself justify intervention. As

Anderson and others have argued, instability from climate variability is a fact of agricultural life for which there is already an array of risk management tools (Trewin *et al* 1992). There are also alternative put-aside strategies, including enterprise diversification, unsubsidised savings and investment strategies, drawdown of off-farm assets and off-farm employment.

4.5 WELFARE PAYMENTS

Generally, Australian farmers (and other self-employed workforce participants) are required to meet both an assets test and an 'availability for work' test before they are able to access general welfare provisions available to other income-earning groups within the community. These tests are designed to ensure that unintended subsidies are avoided and that applicants meet 'mutual obligations' to the rest of the community— that is, those paying for the benefits that are provided.

Under current drought policy, access to welfare payments is triggered by the declaration of exceptional circumstances as discussed in Botterill in this volume and the Exceptional Circumstances Handbook (AFFA 2003).

These arrangements directly address welfare concerns and there can be little doubt that they are popular amongst farmers and enjoy relatively strong public support (Botterill and White *et al*, this volume). But it will be obvious from our previous discussion that there are potential concerns about the trade-off between efficiency and equity aspects of government intervention.

In reality, the welfare payments associated with exceptional circumstances are *not* components of the general welfare system. They are *special* short-term, event-specific payments to a relatively small segment of the self-employed sector. They are therefore subject to the same kinds of reservations that may be held about the use of short-term assistance generally and can, in this sense, be regarded as quasi, industry-specific subsidies with many of the same attributes as, say, interest rate subsidies. They can distort resource allocation within and between sectors, hinder adjustment and contribute towards adverse impacts on the environment. Moreover, they are inequitable in that they are only available to drought-affected farmers (many of whom could be better off than nondrought-affected farmers) and not to other self-employed sectors (which may also be affected adversely by climate variability).

There can be few doubts that such payments do address real welfare-related concerns. But it is doubtful that addressing those concerns through measures that are likely to impede adjustment will result in improved community welfare overall. A number of alternative 'second best' options have been suggested to address this trade-off. Freebairn (Freebairn 2003) has suggested review of access by all self-employed sectors to the welfare system. This should be done. But it will involve highly complex considerations, given the sheer numbers and varying circumstances of those who would regard themselves as self-employed, and could risk creating even greater efficiency concerns. Another option which has been suggested as a replacement for drought support generally is implementation of a modified higher education contribution scheme

(HECS) involving loans with repayments delayed until income levels have been restored to a predetermined level (Botterill and Chapman 2002). Such a scheme could lead to improved 'second best' outcomes but it risks similar moral hazards to existing arrangements and begs the question of 'why just farmers?'

4.6 OBJECTIVES

We turn finally in this assessment to a consideration of whether current intervention supports its policy objectives. An initial observation concerning objectives is that the broad set of objectives which have been adopted for drought policy intervention since the inception of the NDP are themselves internally inconsistent. Moreover, that inconsistency has been accentuated by the nature of the intervention measures chosen. It will be clear, for instance, that there is significant potential for conflict between encouraging self-reliance on the one hand and providing income support for those farmers unable to meet living expenses on the other. Similarly, there is conflict between the latter objective and that of not impeding economic adjustment. The provision of short-term subsidies, even those restricted to exceptional circumstances, creates potential moral hazards for policy objectives associated with preparedness and self-reliance. Indeed, where the behavioural objective of increased self-reliance is not actually met, it is arguable that 'band aid' policies, though not optimal, may be more efficient than those which seek to encourage drought risk preparedness (Simmons 1993).

These ambiguities have been underscored by the policy drift which has characterised drought policy making since the introduction of the NDP in the early 1990s (Botterill 2003d and White *et al*, this volume). Ambiguity of this kind is unlikely to enhance policy effectiveness, which requires not only a clear and consistent set of policy objectives and supporting intervention measures, but also a clear and consistent focus on the policy problem being addressed (for example, see Howlett and Ramesh 1995). The DPRTF report (1990), which was largely responsible for the underlying early rhetoric of the NDP, seems to have been in no doubt that the problem related to the realities of drought as a normal feature of climate variability and the need to manage for it. That focus has been significantly diluted—arguably to the point where drought policy has retreated to pre-NDP rhetoric. This is captured nicely by the Exceptional Circumstances Handbook which states that 'The rationale for providing EC support is to ensure that farmers with long-term prospects of viability will not be forced to leave the land due to short-term events beyond their control' (AFFA 2003).

5. Suggestions for policy improvement

5.1 EXCEPTIONALITY

It was initially envisaged that event-oriented drought support would be provided only in 'exceptional circumstances'. State governments reserved the right to offer their own forms of drought support and in practice they have not always observed the requirement for exceptional circumstances to be declared. Moreover, the national government, with state government support, has watered down the definition of what would constitute an exceptional circumstance over time (O'Meagher *et al* 1998 and White *et al*, this volume).

The original concept of exceptionality focussed on circumstances that were regarded as those beyond which could reasonably be expected to be factored into normal risk management regimes. The one in twenty to twenty-five year events which have been regarded as the definitional centre of the concept was based on the rough and ready concept of a once in a generation event (White and Karssies 1997).

On reflection, for a country where climatic extremes are such a well-recognised fact of agricultural life, this seems a curious policy concept to have adopted. It was of course driven not by rational policy making focussed on natural realities and constraints but rather by political constraints and what was judged at the time to have been acceptable (Botterill 2003d). It is after all an artificial construct, and though a workable definition can be (and arguably was, at least initially) constructed, it has for the reasons outlined by White *et al* in this volume become an all too frequent event which seems to bear little relation to its original intent.

The outcome, as outlined above, has been the undermining of good intentions and the perpetuation of inefficient and inequitable drought policy intervention outcomes.

5.2 AN ALTERNATIVE APPROACH

One approach to this situation would be to revisit the definition of exceptional circumstances and tighten it up to at least reflect the original intention of 'once in a generation' type events. This approach would not address the fundamental problem of moral hazard which confronts current policy intervention. Nor is it likely to be politically feasible.

A more robust approach may be to accept that, like drought itself, extremity is also a normal feature of the Australian landscape to be factored into risk management. Policy intervention would then deal with the fundamental challenge of reducing uncertainty and risk in ways that do not compromise efficiency and equity considerations and are more likely to contribute to overall community welfare.

This approach would dispense with the concept of exceptionality altogether. Accordingly, all short-term, event-oriented support—interest, transaction-based and savings subsidies—would also be dispensed with. Emphasis would instead be focussed on research and extension services which address clear and demonstrable market failures and which are, to the extent appropriate and possible, exposed to cost recovery and contestability principles to help ensure that private sector innovation and intermediation are not impeded. Welfare concerns would be addressed not by special short-term, event-oriented intervention but by the general provisions of the existing welfare safety net.

Such an approach would be consistent with the overall thrust of microeconomic reform in Australia and would remove a source of friction undermining improved welfare for the Australian community.

CHAPTER 11: DROUGHT POLICY AND PREPAREDNESS: THE AUSTRALIAN EXPERIENCE IN AN INTERNATIONAL CONTEXT[*]

DONALD A. WILHITE
National Drought Mitigation Center and International Drought Information Center, University of Nebraska–Lincoln, 239 LW Chase Hall, Lincoln, Nebraska 68583, USA

1. Introduction

Drought is a normal part of the climate for virtually all climate regimes. It is a complex, slow-onset phenomenon that affects more people than any other natural hazard and results in serious economic, social, and environmental impacts. Drought affects both developing and developed countries, but in substantially different ways (Wilhite 2000b, pp3-4). The impacts of drought are often an indicator of nonsustainable land and water management practices, and drought assistance or relief provided by governments and donors can encourage land managers and others to continue these practices. This often results in a greater dependence on government and a decline in self-reliance.

Many people consider drought to be largely a natural or physical event. In reality, drought, like other natural hazards, has both a natural and a social component. The risk associated with drought for any region is a product of both the region's exposure to the event and the vulnerability of society to the event. Exposure to drought varies regionally and there is little, if anything, we can do to alter its occurrence. The natural event, commonly referred to as meteorological drought, is a result of the occurrence of persistent large-scale disruptions in the global circulation pattern of the atmosphere that result in significant regional deficiencies of precipitation over an extended period of time.

As vulnerability to drought has increased globally, greater attention has been directed to reducing risks associated with its occurrence through the introduction of planning to improve operational capabilities (for example, prediction capabilities, monitoring and early warning systems, building institutional capacity, education and training) and other mitigation measures that are aimed at reducing drought impacts. Typically, when a natural hazard event and resultant disaster has occurred, governments and donors have followed with impact assessment, response, recovery, and reconstruction activities to return the region or locality to a pre-disaster state. Historically, little attention has been given to preparedness, mitigation, and prediction/early warning actions (that is, risk management) that could reduce future impacts and lessen the need for government intervention in the future. Because of this emphasis on crisis management, many societies have generally moved from one disaster to another with little, if any, reduction

---

[*]Material in this chapter was first published in *Beyond Drought: People, Policy and Perspectives;* Linda Courtenay Botterill and Melanie Fisher (eds). CSIRO PUBLISHING, Melbourne 2003. Reproduced by permission of the Publisher.

in risk. In addition, in drought-prone regions, another drought event is likely to occur
before the region fully recovers from the last event.

Vulnerability is determined by social factors. As population increases, so does pressure
on natural resources. An increase in the number of people also suggests that more
people will live in climatically marginal areas that will have greater exposure to
drought. Population is also migrating from humid, water-surplus climates to more arid,
water-deficient climates and from rural to urban settings for many locations.
Urbanisation is placing more pressure on limited water supplies and the capacity of
water supply systems to deliver that water to users, especially during periods of peak
demand. An increasingly urbanised population is also increasing conflict between
agricultural and urban water users, a trend that will only be exacerbated in the future.
Increasingly sophisticated technology decreases our vulnerability to drought in some
instances while increasing it in others. Greater awareness of our environment and the
need to preserve and restore environmental quality is placing greater pressure on all of
us to be better stewards of natural and biological resources. All of these factors
emphasise that our vulnerability to drought is continually changing and who is most at
risk from these changes must be evaluated. We should expect the impacts of drought in
the future to be different, more complex, and more significant for some economic
sectors, population groups, and regions. Improving drought management implies an
attempt to use natural resources in a more sustainable manner. This will require a
partnership between individuals and government.

This chapter will concentrate on three principal areas. First, progress in drought
planning and preparedness is discussed from an international perspective. This will be
followed by three case studies—the United States, sub-Saharan Africa and Australia.
The latter will necessarily be brief in light of the earlier chapters in this book. The
chapter will conclude with some observations about progress in implementing drought
preparedness and risk management approaches, including current attempts to establish a
global network aimed at improving levels of drought preparedness within and between
regions.

2. Drought policy and preparedness: overview

Although there has been considerable discussion regarding the adoption of risk-based
drought policies and preparedness plans globally, Australia is one of the few countries
that have actually implemented national programs or strategies. There are four key
components in an effective drought risk reduction strategy (O'Meagher *et al* 2000,
p115). These are the availability of timely and reliable information on which to base
decisions; policies and institutional arrangements that encourage assessment,
communication, and application of that information; a suite of appropriate risk
management measures for decision makers; and actions by decision makers that are
effective and consistent.

Article 10 of the UN Convention to Combat Desertification (UNCCD) states that
national action programs should be established to 'identify the factors contributing to

desertification and practical measures necessary to combat desertification and mitigate the effects of drought' (UNCCD 1999, p14). In the past ten years there has been considerable recognition by governments of the need to develop drought preparedness plans and policies to reduce the impacts of drought. Unfortunately, progress in drought preparedness during the last decade has been slow because many nations lack the institutional capacity and human and financial resources necessary to develop comprehensive drought plans and policies. Recent commitments by governments and international organisations combined with new drought monitoring technologies and planning and mitigation methodologies are cause, however, for optimism. The challenge is the implementation of these new policies, methodologies, and technologies. For example, at a meeting of ministerial delegations and representatives of donor organisations for the West Asian and North African countries on opportunities for sustainable investment in rainfed areas held in 2001, the importance of developing and implementing appropriate drought policies and plans was emphasised as an urgent need (Rabat Declaration 2001, p1). Adopting a regional approach to drought management and preparedness was identified as critical to this region, allowing governments that possess experience with drought policies and preparedness to share it with others through regional and global networks.

Drought planning is an integral part of drought policy. The objectives of drought planning will, of course, vary between countries and should reflect unique physical, environmental, socioeconomic, and political characteristics. A generic set of planning objectives has been developed that could be considered as part of a national, state/provincial, or regional planning effort (Wilhite, Hayes *et al* 2000, p697). These planning objectives have been followed or modified by numerous governments at various levels in the United States and elsewhere since the ten-step drought planning process (Wilhite 1991, p29) was originally developed. For example, the process has been followed in Brazil, Cyprus, and Morocco and will likely be applied in many other countries, as drought preparedness becomes a more common practice. These objectives are set out below.

- Collect, analyse, and disseminate drought-related information in a timely and systematic manner.
- Establish criteria for declaring drought and triggering various mitigation and response activities.
- Provide an organisational structure that assures information flow between and within levels of government, as well as with non-governmental organisations, and define the duties and responsibilities of all agencies with respect to drought.
- Maintain a current inventory of drought assistance and mitigation programs used in assessing and responding to drought emergencies, and provide a set of appropriate action recommendations.
- Identify drought-prone areas and vulnerable sectors, population groups, and environments.
- Identify mitigation actions that can be taken to address vulnerabilities and reduce drought impacts.

- Provide a mechanism to ensure timely and accurate assessment of drought's impacts on agriculture, livestock production, industry, municipalities, wildlife, health, and other areas, as well as specific population groups.
- Keep the public informed of current conditions and mitigation and response actions by providing accurate, timely information to media in print and electronic form.
- Establish and pursue a strategy to remove obstacles to the equitable allocation of water during shortages and provide incentives to encourage water conservation.
- Establish a set of procedures to continually evaluate and exercise or test the plan and periodically revise the plan so it will remain responsive to the needs of the people and government ministries.

Drought plans in which mitigation is a key element should have three principal components: monitoring, early warning, and prediction; risk and impact assessment; and mitigation and response. A description of each of these components follows.

3. Drought monitoring, early warning, and prediction

Effective drought early warning systems are an integral part of efforts worldwide to improve drought preparedness. Timely and reliable data and information must be the cornerstone of effective drought policies and plans. Monitoring drought presents some unique challenges because of drought's characteristics. In addition, several types of drought exist, and the factors or parameters that define it will differ from one type to another. For example, meteorological drought is principally defined by a deficiency of precipitation from expected or 'normal' over an extended period of time, while agricultural drought is best characterised by deficiencies in soil moisture. This parameter is a critical factor in defining crop production potential. Hydrological drought, on the other hand, is best defined by deficiencies in surface and subsurface water supplies (that is, reservoir, lake, and ground water levels; stream flow; and snowpack), and its impacts generally lag the occurrence of meteorological and agricultural drought. These types of drought may coexist or may occur separately.

An expert group meeting on early warning systems for drought preparedness, sponsored by the World Meteorological Organisation and others, recently examined the status, shortcomings, and needs of drought early warning systems, and made recommendations on how these systems can help in achieving a greater level of drought preparedness (Wilhite, Sivakumar *et al* 2000, p177) This meeting was organised as part of the World Meteorological Organisation's contribution to the UNCCD meeting in Bonn, Germany, in December 2000. The proceedings of this meeting documented recent efforts in drought early warning systems in countries such as Brazil, China, Hungary, India, Nigeria, South Africa, and the United States, but also noted the activities of regional drought monitoring centres in eastern and southern Africa and efforts in West Asia and North Africa. Shortcomings of current drought early warning systems were noted in the following areas:

- *Data networks*—inadequate density and data quality of meteorological and hydrological networks and lack of data networks on all major climate and water supply parameters;
- *Data sharing*—inadequate data sharing between government agencies and the high cost of data limit the application of data in drought preparedness, mitigation, and response;
- *Early warning system products*—data and information products are often not user friendly and users are often not trained in the application of this information to decision making;
- *Drought forecasts*—unreliable seasonal forecasts and the lack of specificity of information provided by forecasts limit the use of this information by farmers and others;
- *Drought monitoring tools*—inadequate indices for detecting the early onset and end of drought, although the Standardised Precipitation Index was cited as an important new monitoring tool to detect the early emergence of drought;
- *Integrated drought/climate monitoring*—drought monitoring systems should be integrated and based on multiple indicators to fully understand drought magnitude, spatial extent, and impacts;
- *Impact assessment methodology*—lack of impact assessment methodology hinders impact estimates and the activation of mitigation and response programs;
- *Delivery systems*—data and information on emerging drought conditions, seasonal forecasts, and other products are often not delivered to users in a timely manner;
- *Global early warning system*—no historical drought database exists and there is no global drought assessment product that is based on one or two key indicators, which could be helpful to international organisations, non-governmental organisations, and others.

Participants of the expert group meeting on drought early warning systems made several recommendations. First, early warning systems should be considered an integral part of drought preparedness and mitigation plans. Second, priority should be given to improving existing observation networks and establishing new meteorological, agricultural, and hydrological networks in support of drought monitoring efforts.

A trend toward establishment of national and regional drought monitoring centres is apparent. For example, the regional drought monitoring centres in eastern and southern Africa have had a significant impact on the collection and dissemination of drought forecasts/outlooks and early warning information to diverse users throughout these regions since their formation a decade ago (Ambenje 2000, p131). The seasonal precipitation outlooks provide users with broad regional patterns several months in advance. During periods with a strong El Niño signal (that is, higher probability of drought conditions in eastern Australia and southern Africa), the value of this information increases significantly for agriculture and other weather-sensitive sectors. Discussions regarding the establishment of other regional drought centres in other regions are ongoing. For example, UNESCO, following an international drought conference in South Africa in September 1999, proposed a regional drought centre with a broader mission. The challenge is to link these activities closely with national drought policy and preparedness efforts in these regions.

4. Risk and impact assessment

Drought impacts cut across many sectors and across normal divisions of responsibility of local, state/provincial, and federal agencies. Wilhite and Vanyarkho have classified these impacts (Wilhite and Vanyarkho 2000, p248). Risk is defined by both the exposure of a location to the drought hazard and the vulnerability of that location to periods of drought-induced water shortages (Blaikie *et al* 1994, p9). Information on drought impacts and their causes is crucial for reducing risk before drought occurs and for appropriate responses during drought. As part of a drought planning process, technical specialists and members of stakeholder groups that understand those economic sectors, social groups, and ecosystems most at risk from drought should undertake risk assessment.

An approach in accomplishing this risk assessment that has been effective in the United States is to create a series of working groups as a part of the drought planning process (Wilhite, Hayes *et al* 2000, p697). These working groups will assess sectors, population groups, regions, and ecosystems most at risk and identify appropriate and reasonable mitigation measures to address these risks. The number of working groups established varies considerably between states. This process has been widely used in the United States. This process is applied through a methodology for assessing and reducing the risks associated with drought. This methodology was completed recently through collaboration between the NDMC and the Western Drought Coordination Council's Mitigation and Response Working Group (Knutson *et al* 1998, p1) and is available on the NDMC's web site at http://drought.unl.edu. This guide focuses on identifying and ranking drought impacts, determining their underlying causes, and choosing actions to address the underlying causes. This methodology can be employed by each of the working groups.

The steps included in this methodology include:

1. Assemble the team. Select stakeholders, government planners, and others with a working knowledge of drought's effects on primary sectors, regions, and people.
2. Evaluate the effects of past droughts. Identify how drought has affected the region, group, or ecosystem. Consult climatological records to determine the 'drought of record,' the worst drought in recorded history, and project what would happen if a similar drought occurred this year or in the future, considering changes in land use, population growth, and development that has taken place since the last drought. The worst single-year drought or the worst sequence of drought years, or both, could define the drought of record.
3. Rank impacts. Determine which of drought's effects are most urgently in need of attention. Various considerations in prioritising these effects include cost, areal extent, trends over time, public opinion, social equity, and the ability of the affected area to recover.
4. Identify underlying causes. Determine those factors that are causing the highest levels of risk for various sectors, regions, and population groups. For example, an unreliable source of water for municipalities in a particular region may explain the impacts that have resulted from recent droughts in that area. To reduce the potential for drought

impacts in the future, it is necessary to understand the underlying environmental, economic, and social causes of these impacts. To do this, drought impacts must be identified and the reason for their occurrence determined.

5. Identify ways to reduce risk. Identify actions that can be taken before drought that will reduce risk. In the example above, taking steps to identify new or alternative sources of water or implementation of a water conservation plan by a municipality at risk could increase resiliency to subsequent episodes of drought.

6. Write a 'to do' list. Choose actions that are likely to be the most feasible, cost-effective, and socially equitable. Implement steps to address these actions through existing government programs or the legislative process.

The choice of specific actions to deal with the underlying causes of drought impacts will depend on the economic resources available and related social values. Typical concerns are associated with cost and technical feasibility, effectiveness, equity, and cultural perspectives. This process has the potential to lead to the identification of effective and appropriate drought risk reduction activities that will reduce long-term drought impacts, rather than *ad hoc* responses or untested mitigation actions that may not effectively reduce the impact of future droughts.

5. Mitigation and response

Mitigation is defined in several ways in the natural hazards literature. Hy and Waugh (1990, p19) referred to mitigation as activities that reduce the degree of long-term risk to human life and property. These actions normally include insurance strategies, the adoption of building codes, land-use management, risk mapping, tax incentives and disincentives, and diversification. Drought is not often directly responsible for loss of life and its impacts are largely non-structural. Therefore, this definition is not appropriate in this case. The previously stated definition for mitigation in this chapter is short- and long-term actions, programs, or policies implemented during and in advance of drought that reduce the degree of risk to human life, property, and productive capacity.

Mitigation needs to focus on a range of levels from micro to macro. Davies (2000, p10) has classified these levels as national, local government, community, and household. Wilhite (1997, p961) has documented mitigation actions employed by states in the United States through a survey conducted in the early 1990s. Certainly, the range of alternatives would be greater if this survey were duplicated today since much of the country has been in severe to extreme drought conditions since 1996. The activities identified were diverse, reflecting regional differences in impacts, legal and institutional constraints, and institutional arrangements associated with drought plans. These actions represent a full range of possible mitigative actions, from monitoring and assessment programs to the development of drought contingency plans. Some of the actions included were adopted by many states, while others may have been adopted only in a single case.

Many of the mitigative programs implemented by states in the US during recent droughts can be characterised as emergency or short-term actions taken to alleviate the crisis at hand, although these actions can be successful, especially if they are part of a preparedness or mitigation plan. Other activities, such as legislative actions, drought plan development, and the development of water conservation and other public awareness programs, are considered actions with a longer-term vision. As states gain more experience assessing and responding to drought, future actions will undoubtedly become more timely and effective and less reactive. Viewed collectively, the mitigative actions of states in response to recent drought conditions are numerous, but most individual state actions were quite narrow. In the future, state drought plans need to address a broader range of mitigative actions, including provisions for expanding the level of intergovernmental coordination. Table 1 is illustrative of the arsenal of mitigation programs and actions available to states.

Table1. Drought-related mitigative actions of state government in response to recent episodes of drought

Category	Specific Action
Assessment programs	Developed criteria or triggers for drought-related actions Developed early warning system, monitoring program Conducted inventories of data availability Established new data collection networks Monitored vulnerable public water suppliers
Legislation/public policy	Prepared position papers for legislature on public policy issues Examined statutes governing water rights for possible modification during water shortages Passed legislation to protect instream flows Passed legislation providing guaranteed low-interest loans to farmers Imposed limits on urban development
Water supply augmentation/development of new supplies	Issued emergency permits for water use Provided pumps and pipes for distribution Proposed and implemented program to rehabilitate reservoirs to operate at design capacity Undertook water supply vulnerability assessments Inventoried self-supplied industrial water users for possible use of their supplies for emergency public water supplies Inventoried and reviewed reservoir operation plans
Public awareness/education program	Organised drought information meetings for the public and the media Implemented water conservation awareness programs Published and distributed pamphlets to individuals, businesses, and municipalities on water conservation techniques and agricultural drought management strategies Organised workshops on special drought-related topics

Category	Specific Action
	Prepared sample ordinances on water conservation for municipalities and domestic rural supplies
	Established drought information centre as a focal point for activities, information, and assistance
Technical assistance on water conservation and other water-related activities	Provided advice on potential new sources of water
	Evaluated water quantity and quality from new sources
	Advised water suppliers on assessing vulnerability of existing supply system
	Recommended that suppliers adopt water conservation measures
Demand reduction/water conservation programs	Established stronger economic incentives for private investment in water conservation
	Encouraged voluntary water conservation
	Improved water use and conveyance efficiencies
	Implemented water metering and leak detection programs
Emergency response programs	Established alert procedures for water quality problems
	Stockpiled supplies of pumps, pipes, water filters, and other equipment
	Established water hauling programs for livestock from reservoirs and other sources
	Compiled list of locations for livestock watering
	Established hay hotline
	Provided funds for improving water systems, developing new systems, and digging wells
	Provided funds for recovery programs for drought and other natural disasters
	Lowered well intakes on reservoirs for rural water supplies
	Extended boat ramps and docks in recreational areas
	Issued emergency surface water irrigation permits from state waters
	Created low-interest loan and aid program for agricultural sector
	Created a drought property tax credit program for farmers
	Established a tuition assistance program to enable farmers to enrol in farm management programs
Water use conflict resolution	Acted to resolve emerging water use conflicts
	Negotiated with irrigators to gain voluntary restrictions on irrigation in areas where domestic wells were likely to be affected
	Established a water banking program
	Clarified state law regarding sale of water
	Clarified state law on changes in water rights
	Suspended water use permits in watersheds with low water levels
	Investigated complaints of irrigation wells interfering with

Category	Specific Action
	domestic wells
Drought contingency plans	Established state-wide contingency plans
	Recommended to water suppliers the development of drought plans
	Evaluated worst-case drought scenarios for possible further actions
	Established natural hazard mitigation council

6. Examples of international experience with drought policy and preparedness

6.1 THE UNITED STATES

In 1995 the Federal Emergency Management Agency estimated average annual losses because of drought in the United States to be US$6-8 billion, more than for any other natural hazard (Federal Emergency Management Agency 1995, p2). Yet the United States has typically been ill-prepared to effectively deal with the consequences of drought. Historically, the approach to drought management has been to react to the impacts of drought by offering relief to affected areas. These emergency response programs can best be characterised as too little and too late. More importantly, drought relief does little if anything to reduce the vulnerability of the affected area to future drought events. Improving drought management will require a new paradigm, one that encourages preparedness and mitigation through the application of the principles of risk management.

There are several critical points to note about drought in the United States. First, drought occurs somewhere in the United States every year. On average, 14% of the nation is affected each year. Second, the percent area affected is highly variable from year to year, but drought years are often clustered, as in the 1930s, 1950s, late 1980s and early 1990s, and late 1990s and early 2000s. Third, the worst year on record in terms of percent area affected was 1934, when about 65% of the country was in severe to extreme drought. More recent severe drought episodes have generally been in the 40% range, as was the case in 2002. Finally, no trend in the area affected is noticeable. However, impacts associated with drought in the country have increased substantially in magnitude and complexity. The implication is that vulnerability to drought is increasing.

Since 1996 widespread and severe drought conditions have occurred throughout the United States and have raised serious concerns about continuing vulnerability to extended periods of drought-induced water shortages because of the complexity and magnitude of impacts. Many parts of the country have experienced several consecutive years of drought during this time period. At this writing, some western states (for example, Montana) are into their sixth consecutive year. Although it is not unusual for multiple drought years to occur in the drier western states, the occurrence of consecutive drought years in the east is unusual. For example, south-eastern states such as Georgia, Florida, and South Carolina experienced from three to five consecutive drought years from the late 1990s to the early 2000s.

Most recently, drought conditions during the period 2000–03 affected large portions of the eastern and western states. Impacts on public water supplies, agriculture, forests, transportation, energy production, recreation and tourism, and the environment (for example, fisheries, soil erosion, incidence of forest and wild fires) have been substantial and have drawn considerable attention from elected officials and the media, providing additional fuel for the growing debate regarding the lack of a national drought policy and a co-ordinated response effort between federal, state, local, and tribal governments.

6.1.1 State-level Drought Planning

There has been a remarkable increase in the number of states with drought plans during the past two decades. In 1982, only three states had drought plans in place. In early 2004, thirty-seven states had developed plans and four states were at various stages of plan development. The growth in the number of states with drought plans suggests an increased concern at that level about the potential impact of extended water shortages and an attempt to address those concerns through planning. The rapid adoption of drought plans by states is also a clear indication of their benefits.

Initially, drought plans largely focused on response efforts; today the trend in the United States is for states to place greater emphasis on mitigation as the fundamental element of a drought plan. An example of mitigation actions identified recently by the state of Georgia is shown in Table 2. Agriculture, municipal and industrial, and water quality, flora, and fauna sectors were used to classify these potential mitigation actions.

Initially, states were slow to develop drought plans because the planning process was unfamiliar. With the development of drought planning models (Wilhite 1991, p29; Wilhite, Hayes *et al* 2000, p697) and the availability of a greater number of drought plans for comparison, drought planning has become a less mysterious process for states. As states initiate the planning process, one of their first actions is to study the drought plans of other states to compare methodology and organisational structure.

Many US states have followed to a considerable degree the planning methodology outlined by Wilhite (1991, p29) and Wilhite, Hayes *et al* (2000, p697) in the development of a plan. Tribal and local governments have also used this methodology. At times, this methodology has been followed unknowingly as some states borrow the organisational structure from adjacent or other states that have employed this methodology.

With the tremendous advances in drought planning at the state level in recent years, it should come as no surprise that states have been extremely frustrated and dissatisfied with the lack of progress at the federal level. Early into the 1995–96 drought, the lack of leadership and coordination at the federal level quickly became obvious and continued with subsequent drought episodes. Recent initiatives toward development of a national drought policy are aimed at reducing or eliminating those frustrations.

Table2. Summary of selected pre-drought strategies included in the Georgia Drought Management Plan (Georgia Department of Natural Resources 2003, pp7-12)

MUNICIPAL AND INDUSTRIAL	AGRICULTURE	WATER QUALITY, FLORA, AND FAUNA
State Actions	*Farmer Irrigation Education*	*State Actions*
Formalise the Drought Response Committee as a means of expediting communications among state, local, and federal agencies and non-governmental entities	Recommend that farmers attend classes in best management practices (BMP) and conservation irrigation, before (i) receiving a permit, (ii) using a new irrigation system, or (iii) irrigating for a coming announced drought season	Encourage all responsible agencies to promote voluntary water conservation through a wide range of activities
Establish a drought communications system between the state and local governments and water systems	Provide continuing education opportunities for farmers	Monitor stream flow and precipitation at selected locations on critical streams
Review the local governments' and water supply providers' conservation and drought contingency plans	Develop electronic database for communicating with water use permit holders	Provide the stream flow and water-quality data in real time for use by drought managers and work with drought managers to optimise information delivery and use
Work with the golf course and turf industry to establish criteria for drought-tolerant golf courses	Encourage development and distribution of information on water efficient irrigation techniques	Evaluate the impact of water withdrawals on flow patterns, and the impact of wastewater discharges on water quality during drought
Encourage water re-use	*Field/Crop Type Management* Encourage the use of more drought resistant crops	Investigate indicators and develop tools to analyse drought impacts for waterways such as coastal ecosystems, thermal refuges such as the Flint River, and trout streams.
Provide water efficiency education for industry and business	Encourage the use of innovative cultivation techniques to reduce crop water use	Improve the agencies' capabilities and resources to monitor land-disturbing activities that might result in erosion and sedimentation

MUNICIPAL AND INDUSTRIAL	AGRICULTURE	WATER QUALITY, FLORA, AND FAUNA
		violations
Conduct voluntary water audits for businesses that use water for production of a product or service	Conduct crop irrigation efficiency studies	Identify funding mechanisms and develop rescue and reintroduction protocols for threatened and endangered species during extreme events
Identify vulnerable water dependent industries, fund research to help determine impacts and improve predictive capabilities	Provide farmers with normal year, real time irrigation, irrigation scheduling, and crop evaporation/transpiration information	Develop and execute an effort to identify pollutant load reduction opportunities by wastewater discharge permit holders
Develop criteria for a voluntary certification program for landscape professionals	Monitor soil moisture and provide real time data to farmers	Develop and execute an effort to identify opportunities for industry to decrease water use during drought periods
Develop and implement a state-wide water conservation program to encourage local and regional conservation measures	***Irrigation Equipment Management*** Encourage the installation of water efficient irrigation technology	Evaluate the impact of water withdrawals on flow regimes and the impact of wastewater discharges on water quality during drought
Develop and implement an incentive program to encourage more efficient use of existing water supplies	Retrofit older irrigation systems with newer and better irrigation technology. Update any system over 10 years old	Develop and promote implementation of sustainable lawn care programs based on selected BMPs and/or integrated pest management practices
Local/Regional Actions Develop and implement a drought management and conservation plan.	Encourage farmers to take advantage of available financial incentives for retrofitting and updating older or less efficient systems.	Encourage protection and restoration of vegetated stream buffers, including incentives for property owners to maintain buffers wider than the minimum required by state law.
Assess and classify drought vulnerability of individual water systems.	Recommend irrigation system efficiency audits every 5 to 7 years.	Provide for protection of recharge areas through measures including land purchase or acquisition of easements.
Define pre-determined	***Government Programs***	Encourage and explore

MUNICIPAL AND INDUSTRIAL	AGRICULTURE	WATER QUALITY, FLORA, AND FAUNA
drought responses, with outdoor watering restrictions being at least as restrictive as the state's minimum requirements.	Improve irrigation permit data to create a high degree of confidence in the information on ownership, location, system type, water source, pump capacity, and acres irrigated for all irrigation systems to determine which watersheds and aquifers will be strongly affected by agricultural water use, especially in droughts.	wild-land fire mitigation measures.
Establish a drought communications system from local governments and water supply systems to the public.	Improve on the agriculture irrigation water measurement and accounting state-wide.	Enhance programs to assist landowners and farmers with outdoor burning.
	Improve communications and cooperation among farmers and relevant state and federal agencies regarding available assistance during drought conditions.	
	Support legislation and efforts to enhance the ability of farmers to secure adequate water supplies during drought conditions.	
	Support legislation and efforts to enhance the ability of farmers to secure adequate water supplies during drought conditions.	

6.1.2 *National Drought Policy*

Calls for action on drought policy and plan development in the United States date back to at least the late 1970s. The growing number of calls for action has resulted primarily from the inability of the federal government to adequately address the spiralling impacts associated with drought through the reactive, crisis management approach. This approach has relied on *ad hoc* interagency committees that are quickly disbanded following termination of the drought event. The lessons of these response efforts have quickly been forgotten and the failures of these efforts are subsequently repeated with the next event.

Several regional and national drought-related initiatives occurred as a result of widespread drought conditions in the United States during the period from 1996 to

1998. These initiatives led to the passing of the National Drought Policy Act of 1998, resulting in the formation of the National Drought Policy Commission (NDPC) to 'provide advice and recommendations on creation of an integrated, co-ordinated Federal policy designed to prepare for and respond to serious drought emergencies.' The NDPC's report, submitted to Congress and the president in May 2000, recommended that the United States establish a national drought policy emphasising preparedness (National Drought Policy Commission 2000, p6). The goals of this policy would be to:

1. Incorporate planning, implementation of plans and proactive mitigation measures, risk management, resource stewardship, environmental considerations, and public education as key elements of an effective national drought policy;
2. Improve collaboration among scientists and managers to enhance observation networks, monitoring, prediction, information delivery, and applied research and to foster public understanding of and preparedness for drought;
3. Develop and incorporate comprehensive insurance and financial strategies into drought preparedness plans;
4. Maintain a safety net of emergency relief that emphasises sound stewardship of natural resources and self-help; and
5. Co-ordinate drought programs and resources effectively, efficiently, and in a customer-oriented manner.

The legacy of the 1996 and subsequent droughts is not likely to be their impacts but rather the policy initiatives that occurred in the post-drought period (Wilhite 2001, p20). These initiatives appear to be changing the way droughts are viewed, and they may change the way droughts are managed in the United States. The real question at this point is whether these changes will result in permanent and substantive modifications in the way government entities deal with drought. The National Drought Preparedness Act of 2003 was introduced in the US Congress in July 2003. The goal of this bill is to develop a national drought policy that emphasises risk management through improved levels of monitoring, preparedness, and mitigation. This bill has strong support from the states and bipartisan support in Congress. Now, more than at any time in the history of drought management in the United States, the country is at a critical crossroads for drought policy. Will it continue down the road of crisis management or move toward risk management?

6.2 PROGRESS IN SUB-SAHARAN AFRICA

In sub-Saharan Africa, drought is a major threat to sustainable livelihoods, in particular in dryland areas of arid and semiarid regions (Glantz 1987, p43). Recent drought events have had serious economic, social, and environmental consequences and have resulted in land degradation, human migrations or relocations, famine, diseases, and loss of human life (UNDP/UNSO 2000, p3). In 1986, approximately 185 million people living in the dryland areas of Africa were at risk and 30 million were immediately threatened (Dinar and Keck 2000, p137). Drought has affected nearly all of the countries in western, eastern, and southern Africa in the past two decades, and in many cases on more than one occasion. These droughts have resulted in a recurring deficiency of food supplies and the need for interventions by governments and international donors to alleviate food shortages to avert major losses of human life. For example, the 1991–92 drought in southern Africa resulted in a deficit of more than 6.7 million tonnes of cereal supplies, which affected more than 20 million people (Dinar and Keck 2000, p138). Past drought response programs have been reactive and have done little, if anything, to reduce the impacts of future droughts.

In 1997, a UNDP/UNSO project was initiated to assess the status of drought preparedness and mitigation activities in selected sub-Saharan African countries (UNDP/UNSO 2000, p3). Three main questions were addressed in this assessment. First, what is the status of drought preparedness (that is, institutional capacity) within each country? Second, what constraints exist with regard to policy and plan development? Third, what are the primary drought policy and planning needs? The conclusions summarised here are drawn from eleven of the most drought-prone southern African countries: Angola, Botswana, Lesotho, Malawi, Mauritius, Mozambique, Namibia, South Africa, Swaziland, Zambia, and Zimbabwe.

Common themes on the current status of drought preparedness and institutional capacity in sub-Saharan Africa included the following:

- There is no permanent government body to deal with drought issues;
- Drought response is often co-ordinated through natural disaster authorities;
- Drought relief is directed toward human relief, protection of key assets, and recovery;
- Post-drought evaluation of response is not usually undertaken;
- Formal drought plans are rare and mainly directed at response actions;
- Drought and famine early warning systems commonly co-exist;
- Vulnerability assessments often exist for sectors, groups, and areas at risk;
- Mitigation actions focus on economic diversification and poverty reduction;
- Drought management is increasingly viewed as part of the development process; and
- Drought policies are usually lacking.

Botswana and South Africa clearly stand apart from the other countries included in this assessment in terms of their experiences and current status of drought planning. Although Botswana does not have an identified drought policy and plan, it has had a long history with various types of drought programs. Drought preparedness planning is part of development planning and institutional structure is well defined, with local involvement at the district level. In South Africa, the National Consultative Drought Forum was established in 1992 and composed of representatives of government, church organisations, trade unions, and NGOs. The Forum led to a shift from an exclusive emphasis on commercial farmers to a more comprehensive program that includes rural farmers, rural poor, and farm workers. Policy changes included greater equity for recipients of assistance. Drought policies have increasingly focused on improving levels of self-reliance, reducing risk in the agricultural sector, and stabilising income. The National Drought Management Committee was established in 1995 with similar structures at the provincial and local levels of government. The primary objectives of this committee were to develop national disaster management policy, propose and review new legislation, promote community participation in disaster management, promote the establishment of an integrated disaster information system, and ensure risk reduction at the national level. In 2002 the South African government was looking at additional drought policy revisions (Monnik 2000, p48).

No drought policy or plan currently exists in Angola, Lesotho, Malawi, Mauritius, Mozambique, Namibia, Swaziland, Zambia, or Zimbabwe, although some infrastructure does exist in most of these countries to respond to drought conditions. This has usually been only on a reactive or *ad hoc* crisis management basis. Two early warning systems are often in place, one focusing on monitoring climate and water supply conditions and the other emphasising issues associated with food security. Vulnerable sectors, peoples, or regions have been identified in many of these countries but mitigation actions and programs have been limited. Response actions are generally a joint effort between government authorities, donors, NGOs, and others. Most of the countries mentioned above have made considerable progress in coordinating and incorporating the capacities of donors and NGOs in drought-related emergency responses. For example, in Swaziland, a consortium of NGOs has been identified to address the needs of vulnerable population groups.

Numerous constraints to drought policy and plan development were identified in the country reports. These included:

- Poor quality of meteorological networks
- Minimal understanding of drought impacts
- Lack of institutional capacity
- Low level of involvement by NGOs in drought management
- Lack of understanding of household vulnerability
- Inadequate financial resources for drought management and human resources development
- Need for expanded extension services
- Inequitable access to land
- Limited coordination between government agencies
- Reduced response/mitigation capability due to lack of drought policy and plan

Future drought policy and planning needs were also identified in the country reports. Many of these needs are aimed at addressing the constraints referred to previously. In many countries it was reported that recommendations on drought policies and specific mitigation actions had been made in government reports or as a result of workshops focused on future drought planning and response needs. In many cases, however, these recommendations have not been implemented. For example, Namibia has developed a series of drought policy recommendations based on the elements of the ten-step drought planning process developed by Wilhite (Wilhite 1991, p29; Wilhite, Hayes *et al* 2000, p697). The goal of the Namibian policy is to develop an efficient, equitable, and sustainable approach to drought management that shifts responsibility from government to the farmer. The tenets of that policy are to (1) ensure household food security is not compromised by drought; (2) encourage and help farmers adopt a self-reliant approach to drought risk; (3) preserve reproductive capacity of the national livestock herd during drought; (4) ensure a continuous supply of potable water to communities and livestock; (5) prevent degradation of the natural resource base; (6) enable rural inhabitants and the agricultural sector to recover quickly following drought; (7) ensure the health status of all Namibians; and (8) finance drought relief programs efficiently by establishing an independent and permanent national drought fund.

Increased interagency coordination and the need to enhance institutional capacity were also considered important. Other needs identified included creation of a permanent national drought fund in support of mitigation and response measures, expanded meteorological networks and more comprehensive early warning systems, improved vulnerability assessments and vulnerability tracking systems, increased community participation and involvement, expanded NGO involvement in drought management, and the development of strategic grain reserves.

As expected, there is a wide range of institutional capacity to respond to drought emergencies in southern Africa. Although some countries have an organisational structure in place to co-ordinate the actions of government at various levels, as well as those of donors and nongovernmental organisations, most have not developed a permanent institutional capacity. One of the common problems with drought and other natural hazards is maintaining interest in planning beyond the relatively short window of opportunity that follows the event, given the on-again, off-again nature of drought. Interest in drought planning quickly wanes in the post-drought period when precipitation conditions have returned to normal or above-normal levels. The challenge is to break this cycle by developing and implementing comprehensive drought preparedness plans that emphasise risk management.

6.3 AUSTRALIA

As outlined in earlier chapters, Australia officially adopted a risk management approach to drought in 1992. This policy included many of the characteristics outlined above with its focus on increased research and development on climate patterns, an emphasis on self-reliance by agricultural producers and the intention to move away from *ad hoc* responses to drought. As illustrated elsewhere in this volume the implementation of the National Drought Policy has not always met its objectives; however, it is a step in the right direction. It also highlights the difficulties governments can face in implementing a preparedness approach to drought, even in comparatively wealthy countries in which drought is a recurring phenomenon.

7. Global drought preparedness network

Because of increasing concern over the escalating impacts of drought and society's inability to effectively respond to these events in the past, developing and developed countries are now placing greater emphasis on the development of national policies and plans that stress the principles of risk management. Global initiatives, such as the UN Convention to Combat Desertification (UNCCD), are emphasising the importance of improving drought early warning systems and seasonal climate forecasts and developing drought preparedness plans.

The National Drought Mitigation Center at the University of Nebraska at Lincoln is working in partnership with the United Nations Secretariat of the International Strategy for Natural Disaster Reduction and other organisations to develop a network of regional networks on drought preparedness and then to link these networks into a Global Drought Preparedness Network (GPDN). Working in cooperation with the UN's Secretariat for the International Strategy for Disaster Reduction, the goal is to promote the concepts of drought preparedness and mitigation in order to build greater institutional capacity to cope with future episodes of drought (ISDR Drought Discussion Group 2003, pp10-12). The GDPN could provide the opportunity for nations and regions to share experiences and lessons learned (successes and failures) through a virtual network of regional networks—for example, information on drought policies, emergency response measures, mitigation actions, planning methodologies, stakeholder involvement, early warning systems, automated meteorological networks, the use of climate indices for assessment and triggers for mitigation and response, impact assessment methodologies, demand reduction/water supply augmentation programs and technologies, and procedures for addressing environmental conflicts.

8. Conclusion

As this book has argued for Australia, there is a need internationally to build awareness of drought as a normal part of climate. It is often considered to be a rare and random event—thus the lack of emphasis on preparedness and mitigation. Improved understanding of the different types of drought and the need for multiple definitions and climatic/water supply indicators that are appropriate to various sectors, applications, and regions is a critical part of this awareness-building process.

A second challenge is to erase misunderstandings about drought and society's capacity to mitigate its effects. Many people consider drought to be purely a physical phenomenon. We may ask, if drought is a natural event, what control do we have over its occurrence and the impacts that result? Drought originates from a deficiency of precipitation over an extended period of time. The frequency or probability of occurrence of these deficiencies varies spatially and represents a location's exposure to the occurrence of drought. Some regions have greater exposure than others, and we do not have the capacity to alter that exposure.

As with other natural hazards, drought has both a physical and a social component. It is the social factors, in combination with our exposure, that determines risk to society. Some of the social factors that determine our vulnerability are level of development, population growth and its changing distribution, demographic characteristics, demands on water and other natural resources, government policies (sustainable versus nonsustainable resource management), technological changes, social behaviour, and trends in environmental awareness and concerns. It is obvious that well-conceived policies, preparedness plans, and mitigation programs can greatly reduce societal vulnerability and therefore the risks associated with drought.

A fourth challenge is to convince policy and other decision makers that investments in mitigation are more cost effective than post-impact assistance or relief programs. Evidence from around the world, although sketchy, illustrates that there is an escalating trend of losses associated with drought in both developing and developed countries. Also, the complexity of impacts is increasing. It seems clear that investments in preparedness and mitigation will pay large dividends in reducing the impacts of drought. A growing number of countries are realising the potential advantages of drought planning. Governments are formulating policies and plans that address many of the deficiencies noted from previous response efforts that were largely reactive. Most of the progress made in drought preparedness and mitigation has been accomplished in the past decade or so. Although the road ahead will be difficult and the learning curve steep, the potential rewards are numerous. The crisis management approach of responding to drought has existed for many decades and is ingrained in our cultures and reflected in our institutions. Movement from crisis to risk management will certainly require a paradigm shift. The victims of drought have become accustomed to government assistance programs. In many instances, these misguided and misdirected government programs and policies have promoted the nonsustainable use of natural resources. Many governments have now come to realise that drought response in the form of emergency assistance programs only reinforces poor or nonsustainable actions and decreases self-reliance.

Internationally, progress in drought preparedness is accelerating as knowledge of drought planning tools becomes more widely known and drought impacts increase in magnitude and complexity. Many regional efforts are underway to provide greater emphasis on drought policies and plans. Recent international and regional drought conferences and workshops are good examples of this growing momentum. As nations continue to build institutional capacity to cope with drought, it is imperative that these lessons learned are shared with others. Working individually, many nations and regions will be unable to improve drought coping capacity. Collectively, working through global and regional partnerships, we can achieve the goal of reducing the magnitude of economic, environmental, and social impacts associated with drought in the twenty-first century.

CHAPTER 12: LESSONS FOR AUSTRALIA AND BEYOND

LINDA COURTENAY BOTTERILL
*National Europe Centre, 1 Liversidge Street (#67C), Australian National University,
ACT 0200, Australia*

1. Introduction

Australia has had its National Drought Policy in place for more than a decade. It is therefore timely to consider the strengths and weaknesses of the policy approach that was adopted in 1992 and to draw some lessons for Australia and other countries considering an integrated policy response to drought. Many of the lessons outlined below apply particularly to industrialised countries in which the farm sector is diminishing in importance, in terms of its contribution to GDP, and in which drought does not result in widespread human disasters such as famine.

In summary, policy makers in Australia in 1992 attempted to align attitudes towards drought with the reality of a highly variable climate. The move from a disaster response to an approach based on self-reliance and risk management was based in a recognition that Australian farmers should expect droughts to occur and should factor drought risk into their business decisions. In economic and policy terms, the recommendations of the Drought Policy Review Task Force which reported in 1990 and the direction of the National Drought Policy announced in 1992 were coherent and logical and would allow the farm sector to operate efficiently and productively within the constraints of the Australian climate. However, drought responses are not only concerned with economic and policy coherence—they are developed in a specific socio-political context. The following section discusses the context of Australia's drought response and highlights some of the tensions which arise between different policy objectives and different values within the Australian community and the problems that have arisen in the implementation of the National Drought Policy. The final section identifies the lessons from which Australian policy makers and their counterparts elsewhere in the world can draw in considering future drought responses.

2. Tensions within the National Drought Policy

The collection of papers in this book attempts to illustrate the range of issues that need to be considered by policy makers if they are to develop an equitable, affordable and rational drought response. There are several perspectives at play. First, drought can be considered literally from the ground up. This is the way Australia's indigenous people managed their available water. As Deborah Rose points out in her chapter, 'people sought to enhance water's capacity to nourish life without seeking radically to alter the water conditions of their country or, cumulatively, of the continent'. Rose describes a way of life in which people are 'of the land' rather than 'on the land'. This is a view of

L.C. Botterill and D.A. Wilhite (eds.), From Disaster Response to Risk Management, 177–183.
© 2005 *Springer. Printed in the Netherlands.*

Australian climate which does not conceptualise climate 'events' such as drought as inherently transgressive—the climate just is.

However, the arrival of Europeans on this continent brought with it the introduction of a form of agriculture developed for a more predictable climate cycle. This type of farming provides the second perspective—that of the hard-working farmer struggling against the elements. Land reforms in the mid-nineteenth century resulted in the development of small-scale family farming and a push to closer settlements which persisted well into the twentieth century. These developments were associated with an agrarian view of agriculture which carried with it moral and identity issues relating to the role of farming as intrinsically valuable and special in comparison with other economic activities. Interestingly this perception of farming as an essential activity persists in industrialised countries, even when farming activity is now only a small contributor to national wealth and food shortages are a remote and unpleasant memory.

This view of agriculture carries with it the iconic image of the hard-working farm family battling the elements and at the whim of a sometimes unfair God. This perspective suggests that government intervention should be in the form of disaster relief aimed at sustaining both the farm business and the farm family during the drought event—an approach that was taken in Australia before 1989. Although sympathetic to the plight of farmers, such a policy is arguably a subsidy to farmers to continue operating in areas which are otherwise unprofitable (Winters 1990). For example, sustaining agricultural production in parts of the Western Division of New South Wales through the provision of drought relief is perhaps little different from supporting the growing of sugar beet in Finland—both activities require government support in order to continue operations.

However, as Stehlik illustrates so clearly in her chapter, the struggling farmer is not a mythical character conjured up by the farm lobby to justify government largesse to businesses which are otherwise capable of sustaining themselves. To many smaller operations, drought is a very real and traumatic experience, threatening business, communities and even marriages. The impact of drought on this group cannot be dismissed by policy makers.

The third perspective is that of the farm as a business operation facing a portfolio of risks, one of which is the unreliability of the Australian climate. Agriculture has adapted over the two hundred plus years that it has been practiced in Australia. Donald notes that

> Many innovations have been concerned with problems of rainfall deficiency. They have included the breeding of cereals of early maturity matched to the rainfall season, the development of fodder plants suited to low or strongly seasonal rainfall, diverse means of water storage for livestock, soil management systems for water conservation, fodder conservation for dry periods and irrigation (Donald 1982, p57).

In essence, the National Drought Policy proposed that this adaptiveness be extended to the adoption of farming practices that were appropriate to the *variability* of the climate as well as its dryness. This view calls on farmers to manage drought risk and is based on an (unspoken) assumption that farmers who cannot sustain their farm business throughout the climate cycle are not viable in the long run. This approach, which informs current policy, suggests government intervention in the form of support for improved risk management skills. Difficulties with the self-reliance philosophy arise because of perceived market failures in the delivery of finance to farmers during drought and the challenges associated with the development of risk management tools such as crop insurance, as outlined in Greg Hertzler's chapter.

These three perspectives highlight the difficulties of developing a drought policy for Australia. In addition to this simplified characterisation of the tensions in Australia's drought response, two further key points need to be made. Australia's agricultural policy approach is perhaps unique in the developed world. Since the 1970s it has increasingly deregulated the farm sector and exposed it to world markets. The prevailing policy approach is informed by neoliberal economics and demonstrates a preference for reduced government intervention in the marketplace. In such an environment, policy makers are keen to avoid subsidies which will keep otherwise non-viable farm operators in business. A second contextual point is that Australia's farming industry is not homogeneous. As a rough rule of thumb, the top 20% of agricultural producers are responsible for 80% of output. Although the vast majority of farm operations are characterised as 'family farms', this description conceals a wide range— from larger-than-family farms run purely as businesses to marginal producers struggling to stay on the land. Although many farmers experiencing recent droughts express the sentiments Stehlik reports in her chapter, there are many others who are privately critical of government drought relief but who are unwilling to openly criticise what they regard as poor farm management (Wahlquist 2003).

The risk management approach to drought constituted a policy paradigm shift. Daniela Stehlik's chapter vividly illustrates that, while this new approach was readily accepted in the rural policy community, it has not been welcomed by many farmers who do not accept its logic and feel abandoned and betrayed by the non-farm community. As Hayman and Cox explain so clearly in their chapter, 'when a policy economist asks a farmer to consider drought as a normal recurring business risk, some interpret this as asking farmers to take the enlarged metaphor of rural suffering as a normal and recurring risk'. Stehlik describes the change in policy direction as a move from the consideration of drought as a collective problem to its framing as an individual concern.

Farmers are not the only members of the community who are ambivalent to the policy change. Perhaps more as a result of ignorance of the policy direction, the broader community continues to react to drought in terms of disaster. This is partly due to the way in which the media presents the issue to its audience and partly due to a residual agrarianism that generates an empathy for farmers in difficulty and stimulates the levels of generosity demonstrated in the two Farm Hand appeals. As Ward points out in his chapter, 'television coverage of drought as natural disaster may still prejudice public understanding of drought'. Even when media organisations appear to take a tough line on farm support in good times they are likely to slip back into agrarian language and sentiments once drought emerges as a public issue.

At the heart of the policy, however, is a recognition that Australia's climate is highly variable so although the implementation of a risk management approach is fraught with difficulties, it is equally difficult to argue that a disaster response is appropriate. Policy makers are therefore faced with balancing the tensions between an approach which is responsive to the needs of farmers (particularly those who are less productive or are working marginal land), the imperatives of achieving environmental sustainability, and the logic of working with the climate as it is and managing its variability along with other business risks.

The problems with the implementation of the National Drought Policy in many ways reflect these tensions. These have been described in a number of the chapters in this collection but can be broadly summarised as definitional and political. The lack of a definition of 'exceptional circumstances' in the implementing legislation and its accompanying material was a major obstacle to the early acceptance of the risk management approach to drought. Given the nature of drought as a slowly emerging phenomenon, the distinctions between a dry spell and a drought and between a 'normal' drought and an exceptional event are far from clear-cut. As described by White et al in their chapter, attempts at developing a scientific definition of an exceptional event have made great progress; however, the declaration process remains highly contested. There is an extensive international literature on the difficulties of arriving at an agreed definition of drought (see for example Dracup et al 1980; Wilhite 2000a; Wilhite and Glantz 1985) and it is generally agreed that there can be no universal definition of drought. Given the size of the Australian continent, important differences between biophysical regions need to be taken into account in determining when an event is exceptional in its scale and impact. The six criteria developed in 1994 were broadly accepted; however, the determination of the threshold levels at which those criteria are met is not straightforward. The National Drought Policy has been subject to a number of changes over the period of its operation. O'Meagher points out that this 'policy drift' has highlighted the ambiguities in the policy approach—ambiguities which mirror the tensions outlined above.

The emphasis on the family farm as the preferred form of agricultural production introduces a further complication into the delivery of drought relief. In spite of recent attempts by governments to describe farmers as 'farm business managers' (Crean 1992, p2412) and references to the 'farm family business' (Anderson 1997a), it remains true of much of family farming in Australia that it is characterized by a 'unity of business and household' (Mauldon and Schapper 1974, 65). This raises the policy question of whether drought relief should be delivered to the farm unit or whether separate arrangements should be made to address the welfare needs of the farm family and the support needs of the farm business. As outlined in Botterill's chapter in this volume, the National Drought Policy initially did not deliver any welfare support other than a loans-related scheme which was linked to farm exit. In 1994, in response to the perception of a growing welfare problem as a result of the drought, the Drought Relief Payment was introduced.   This scheme and its successor, the Exceptional

Circumstances Relief Payment, greatly increased the desirability of obtaining an exceptional circumstances declaration as it augmented interest rate subsidies for productive farmers with an income support scheme for all farmers in an EC area, irrespective of the health of their businesses.

The income support element of the drought policy response is arguably the major obstacle to the acceptance of a national drought policy based on risk management and self-reliance. The focus of media reactions to drought is on barefoot children, dying sheep and dusty paddocks—with an emphasis on the day-to-day needs of the farm family. If these were to be addressed equitably, political pressure for government intervention during drought is likely to diminish. Australia's general social welfare safety net is primarily focused on wage and salary owners and the asset-rich, income-poor status of farm families can exclude them from the welfare support which is available to other members of the community. Poverty on Australia's farms has not been studied in any detail since the mid-1970s when two studies were undertaken as part of a larger inquiry into poverty in Australia (Musgrave *et al* 1975; Vincent *et al* 1975). These studies concluded that farm poverty was linked to slower-than-optimal structural adjustment in agriculture and recommended that the appropriate response was structural adjustment support to the farm sector.

Since that time, programs have been developed without a detailed study of the nature, extent or causes of farm poverty, based on largely untested assumptions. These assumptions have not always proved to be correct (Botterill 2001). Governments have received conflicting advice about whether the welfare component of drought relief should be delivered through structural adjustment programs or through separate schemes. In its 1990 Report, the Drought Policy Review Task Force suggested that the government should treat the farm as a single entity by recommending that '[t]he income support needs of rural families in severe financial difficulties are appropriately addressed through the Rural Adjustment Scheme' (DPRTF 1990, 27). Only two years later, the consultants reviewing the Rural Adjustment Scheme gave the opposite advice (Synapse Consulting (Aust) Pty Ltd 1992, ix). More recently, the 1997 review of the RAS stated that

> Welfare assistance should not be delivered through instruments that assist businesses. Such an approach confuses the objective of the intervention, does not effectively target the welfare problem and distorts market signals to farm businesses receiving assistance. (McColl *et al* 1997, 38)

Until the issue of farm poverty is examined empirically and an equitable response developed through the social welfare system, issues of income support during drought will continue to dominate the political debate surrounding drought relief.

3. Opportunities for lesson drawing and possible future directions

As one of the leaders in the development of a comprehensive national drought policy framework, Australia provides the opportunity for other countries to benefit from its experience. There is a growing literature on the nature of lesson drawing in public policy and the advantages for policy makers in taking the shortcuts afforded by the experience of others in tackling common problems in different policy settings (see for example Dolowitz and Marsh 2000; May 1992; Rose 1991; Rose 1993; Schneider and Ingram 1988). Much of this lesson drawing takes place across national boundaries but within the same broad policy arena. Learning from others' experience can assist busy policy makers who are increasingly facing resource constraints which limit their capacity to search for alternative policy approaches. In addition, it is generally acknowledged that human actors lack the cognitive capacity and the time and physical resources to undertake a rational comprehensive approach to problem solving (Albaek 1995, p83; John 1998, p125; Lindblom 1959; Simon 1953). In other words, the expectation that policy is developed through the identification of objectives, the collection of information, the development of alternatives and the ranking of options is an ideal rarely attained in the real world. Policy development is much more likely to involve 'satisficing' (Simon 1953) and incrementalism (Lindblom 1959). Satisficing is a term coined by Herbert Simon which describes the practice of ceasing the search for a solution once the first satisfactory answer has been found—it may not be the optimum outcome but it will suffice. With March, Simon argued that

> *Most human decision-making, whether individual or organizational, is concerned with the discovery and selection of satisfactory alternatives: only in exceptional cases is it concerned with the discovery and selection of optimal alternatives.* (March and Simon 1958, 141—italics in original)

Drawing lessons from the experience of others, notably within the same epistemic community, provides a means for developing better policy within the constraints faced by the decision maker. However, 'pinching ideas' (Schneider and Ingram 1988) from other jurisdictions is not always successful. In this context, May distinguishes between lesson drawing and mimicry (May 1992, 333).

One of the objectives of the present volume is to highlight the successes and failures of the Australian approach and to provide lessons from which policy makers can draw. There can be little doubt that the appropriate response to drought in Australia, and elsewhere in the developed world, is to recognise that it is a normal part of climate. Against this background, policy communities are then faced with developing responses that increase public awareness of the realities of climate and improve our understanding of climate patterns to contribute to better informed risk management decisions. As Janette Lindesay points out in her chapter, the incidence of severe drought in Australia is likely to increase as a result of climate change, so a sound coherent policy response based on the realities of climate will become increasingly important.

During 2004, the Australian government will be considering a major review of drought policy. Policy makers will once again be faced with the tensions between the economic, social and environmental impacts of drought and will be performing the balancing act between the demands of competing interests. It has been argued elsewhere that there is no single, objectively correct drought policy waiting to be found but that any policy will involve trade-offs between competing objectives (Botterill 2003b). An important lesson to be learnt from the Australian experience is that the policy process needs to be informed by multi-disciplinary analysis which allows for the consideration of different and often competing perspectives.   This is not an easy task, and it is hoped that the views offered in this collection will provide valuable background for policy makers in Australia and elsewhere faced with the development of a coherent drought policy.

REFERENCES

AAC (Australian Agricultural Council) (1992) *Record and Resolutions: 137th Meeting, Fremantle, 14 February 1992*, Commonwealth of Australia

ABARE (Australian Bureau of Agricultural and Resource Economics) (2003) *Winter rains boost crop prospects* http://www.abareconomics.com/newwebdesign/internal.html

ACANZ (Agricultural Council of Australia and New Zealand) (1992) *Record and Resolutions: 138th Meeting, Mackay 24 July 1992*, Commonwealth of Australia

AFFA (2003) *Information Handbook, Exceptional Circumstances Assistance.* Guide to the policy and assistance provided under Exceptional Circumstances (EC) Canberra, Department of Agriculture, Fisheries and Forestry, September 2003
http://www.affa.gov.au/corporate_docs/publications/pdf/innovation/drought/ec_handbook_sept03.pdf

Aitkin, D (1985) "Countrymindedness—The spread of an idea" *Australian Cultural History* (4): 34-41

Albaek, E (1995) "Between knowledge and power: Utilization of social science in public policy making" *Policy Sciences* 28: 79-100

Allan, R and R L Heathcote (1987) "The 1982-83 drought in Australia" in M H Glantz, R Katz and M Krenz (Eds), *The Societal Impacts Associated with the 1982-93 Worldwide Climate Anomalies* Boulder, Colorado, National Center for Atmospheric Research: 18-23

Allan, R J and J A Lindesay (1998) "Past climates of Australasia" in J E Hobbs, J A Lindesay and H A Bridgman (Eds), *Climates of the Southern Continents: Present, Past and Future* Chichester, John Wiley & Sons: 207-247

Allan, R J, J A Lindesay and D Parker (1996) *The El Niño Southern Oscillation and Climatic Variability* Melbourne, CSIRO Publishing

Alston, M (2002) "Social Capital in Rural Australia" *Rural Society* 12(2): 93-104

Ambenje, P G (2000) "Regional Drought Monitoring Centers—The Case of Eastern and Southern Africa" in D A Wilhite, M V K Sivakumar and D A Wood (Eds), *Early Warning Systems for Drought Preparedness and Drought Management, Proceedings of an Expert Group Meeting Lisbon, Portugal, 5-7 September* Geneva, Switzerland, World Meteorological Organization: 131-136

Amery, R, the Hon MP (2002) *Howard Government Paying Little and Wanting to Pay Less* Media Release by the NSW Minister for Agriculture, 11 October 2002

Anderson, J, the Hon MP (1997a) *Federal Government gives farm sector 'AAA' rating* Media Release by Minister for Primary Industries and Energy, 14 September

Anderson, J, the Hon MP (1997b) *Farm Household Support Amendment (Restart and Exceptional Circumstances) Bill: Second Reading Speech* House of Representatives Hansard 2 October 1997

Anderson, J, the Hon MP (2002) *Question without notice: Drought* House of Representatives Hansard 23 September 2002

Anderson, J R (1979) "Impacts of Climate Variability in Australian agriculture: A Review" *Review of Marketing and Agricultural Economics* 47(3): 147-77

Anderson, J R (1994) "Risk Management in Australian Agriculture: An Overview" in R Powell (Ed.) *Risk Management in Australian Agriculture*, UNE

Anderson, J R and J L Dillon (1992) *Risk analysis in dryland farming systems* Rome, FAO

186

Anderson, K and R Garnaut (1987) *Australian Protectionism: Extent, Causes and Effects* Sydney, Allen and Unwin

Andrews, J (1966) "The emergence of the wheat belt in south-eastern Australia to 1930" in J Andrews (Ed.) *Frontiers and men* Melbourne, Cheshire

ARMCANZ (Agriculture and Resource Management Council of Australia and New Zealand) (1994a) *Record and Resolutions: Fourth Meeting, Adelaide, 28 October 1994*, Commonwealth of Australia

ARMCANZ (Agriculture and Resource Management Council of Australia and New Zealand) (1994b) *Record and Resolutions: Third Meeting, Canberra, 23 September 1994*, Commonwealth of Australia

ARMCANZ (Agriculture and Resource Management Council of Australia and New Zealand) (1995) *Record and Resolutions: Sixth Meeting, Perth, 18 August 1995*, Commonwealth of Australia

ARMCANZ (Agriculture and Resource Management Council of Australia and New Zealand) (1997) *Record and Resolutions: Ninth Meeting, Melbourne 28 February 1997*, Canberra, Commonwealth of Australia

ARMCANZ (Agriculture and Resource Management Council of Australia and New Zealand) (1999a) *Record and Resolutions: Fifteenth Meeting Adelaide 5 March 1999*, Canberra, Commonwealth of Australia

ARMCANZ (Agriculture and Resource Management Council of Australia and New Zealand) (1999b) *Record and Resolutions: Sixteenth Meeting Sydney 6 August 1999*, Canberra, Commonwealth of Australia

ARMCANZ (Agriculture and Resource Management Council of Australia and New Zealand) (2001a) *Record and Resolutions: Nineteenth Meeting Wellington 9 March 2001*, Canberra, Commonwealth of Australia

ARMCANZ (Agriculture and Resource Management Council of Australia and New Zealand) (2001b) *Record and Resolutions: Twenty first Meeting Darwin 17 August 2001*, Canberra, Commonwealth of Australia 2001

Arrow, K J (1996) "The Theory of Risk-Bearing: Small and Great Risks" *Journal of Risk and Uncertainty* 12: 103-111

Arthur, J (1999) "Dictionaries of the default country" *Lingua Franca, ABC Radio National*

Aufe der Heide, E (1989) *Disaster Response. Principles of Preparation and Coordination* St Louis, C.V. Mosby

AUSLIG (1992) *The Ausmap Atlas of Australia* Cambridge, Cambridge University Press

Australian Broadcasting Corporation (2002) *Drought Money* 2002 www.abc.net.au/rural/qld/stories/s551689.htm

Australian National Audit Office (2003) *Administration of Three Key Components of the Agriculture— Advancing Australia (AAA) Package* Canberra

Baldock, I (2003) *A retailer's tale* Paper presented to the National Party Water Forum, The Brisbane Institute http://www.brisinst.org.au/resources/brisbane_institute_baldock_drought.html

Bardsley, P, A Abbey and S Davenport (1984) "The Economics of Insuring Crops against Drought" *Australian Journal of Agricultural Economics* 28: 1-14

Bardsley, P and M Harris (1987) "An approach to the econometric estimation of attitudes to risk in agriculture" *Australian Journal of Agricultural Economics* 31: 112-126

Barker, W R and P J M Greenslade (Eds) (1982) *Evolution of the Flora and Fauna of Arid Australia* Adelaide, Peacock Publications

Beadle, N C W (1981) *Vegetation of Australia* Cambridge, Cambridge University Press

Beck, U (1992) *Risk society: Towards a new modernity* Theory, culture & society. London, Sage

Beckett, J (1958) "Marginal Men: A Study of Two Half-caste Aborigines" *Oceania* 29(2): 91-108

Bedo, D (1997) *Rainfall decile analysis and drought exceptional circumstances* Proceedings of a workshop on Indicators of Drought Exceptional Circumstances, 30 September—1 October 1996 Canberra, Bureau of Resource Sciences

Benson, C and E Clay (2000) "The Economic Dimensions of Drought in Sub-Sahara Africa" in D A Wilhite (Ed.) *Drought: A Global Assessment* London, Routledge Publishers Volume 1:287-311

Bensen, J (2002) "Letter to the editor" *Sydney Morning Herald* 12th July 2002

Benthall, J (1993) *Disasters, Relief and the Media* London, Tauris

Berkowitz, D (1997) "Non-routine news and news work" in D Berkowitz (Ed.) *Social Meanings of News* Thousand Oaks, Sage

Bernstein, P L (1996) *Against the Gods: The Remarkable Story of Risk* New York, John Wiley

Beynon, N, D Karoly and D White (2000) *Climate Variability in Agriculture R&D Program. Mid-term Review* Land and Water Resources Research and Development Corporation

Bigge, J T (1966) *Report on Agriculture and Trade in NSW* Australiana Facsimile Editions No 70 Adelaide, Libraries Board of South Australia

Birkland, T A (1996) "Natural disasters as focusing events: Policy communities and political response" *International Journal of Mass Emergencies and Disasters* 14(2): 221-243

Blaikie, P, T Cannon, I Davis and B Wisner (1994) *At Risk: Natural Hazards, People's Vulnerability, and Disasters* London, Routledge

Blench, R and Z Marriage (1999) *Drought and livestock in semi-arid Africa and southwest Asia* Overseas Development Institute Working Paper 117 London

Bolduc, J (1987) "Natural disasters in developing countries: Myths and the role of the media" *Emergency Preparedness Digest (Canada)* 14(3): 12

BOM (Bureau of Meteorology) (2002) *Rainfall deficiencies worsen following dry October* Drought Statement—Issued 1st November 2002 Melbourne

BOM (Bureau of Meteorology) (2003) *Above average falls likely in tropical Australia* National Seasonal Rainfall Outlook: Probabilities for November 2003 to January 2004 Melbourne 16 October 2003

Bond, G E and B S Wonder (1980) "Risk attitudes amongst Australian farmers" *Australian Journal of Agricultural Economics* 24: 16-35

Botterill, L C (2001) "Rural Policy Assumptions and Policy Failure: The Case of the Re-establishment Grant" *Australian Journal of Public Administration* 60(4): 13-20

Botterill, L (2003a) *Government Responses to Farm Poverty 1989-1998: The Policy Development Process* Rural Industries Research and Development Corporation RIRDC Publication No 02/163 Canberra

Botterill, L C (2003b) "Beyond drought in Australia: The way forward" in L C Botterill and M Fisher (Eds), *Beyond Drought—People, Policy and Perspectives* Melbourne, CSIRO Publishing: 197-207

Botterill, L C (2003c) "Government Responses to Drought in Australia" in L C Botterill and M Fisher (Eds), *Beyond Drought—People, Policy and Perspectives* Melbourne, CSIRO Publishing: 49-65

188

Botterill, L C (2003d) "Uncertain Climate: The Recent History of Drought Policy in Australia" *Australian Journal of Politics and History* 49(1): 61-74

Botterill, L and B Chapman (2002) "Developing Equitable and Affordable Government Responses to Drought in Australia" *Centre for Economic Policy Research Discussion Paper* Discussion Paper No 455

Botterill, L C and M Fisher (Eds) (2003) *Beyond Drought: People, Policy and Perspectives* Melbourne, CSIRO Publishing

Bowman, P J, G M McKeon and D H White (1995) "The impact of long range climate forecasting on the performance of sheep flocks in Victoria" *Australian Journal of Agricultural Research* 46: 687-702

Bradley, J (1988) *Yanyuwa country: The Yanyuwa people of Borroloola tell the history of their land* Richmond, Greenhouse Publications

Brinkley, T R, G P Laughlin, M F Hutchinson, and K Rananunga (in press) "GROWEST PLUS—A tool for rapid assessment of seasonal growth for environmental planning and assessment" *Environmental Modelling and Software*

Brook, K (1996) *Development of a National Drought Alert Strategic Information System* Final report on QPI 20 to Land and Water Resources Research and Development Corporation Canberra, LWRRDC

Brownhill, S D (1994a) *Drought Relief Payment Bill 1994: Second Reading Debate* Senate Hansard Canberra 13 October

Brownhill, S D (1994b) *Matters of Public Importance: Drought* Senate Hansard 23 August 1994

Bryant, L (1992) "Social Aspects of the Farm Financial Crisis" in G Lawrence, F Vanclay and B Furze (Eds), *Agriculture, Environment and Society* South Melbourne, Macmillan

Bullen, F (1993) "Plant cover types determined from growth phenology" *ERINYES no. 19, Newsletter of the Environmental Resources Information Network*: 12

Burdon, A (1995) "Dry Paddocks, Damp Policies: Drought Assistance Strategies and Their Effectiveness" in Parliamentary Research Service (Ed.) *Australian rural policy papers 1990-95* Canberra, AGPS: 67-160

Bureau of Meteorology (2003) *Climate of the 20th Century*
http://www.bom.gov.au/lam/climate/levelthree/c20thc

Butlin, N G, A Barnard and J J Pincus (1982) *Government and Capitalism: Public and Private Choice in Twentieth Century Australia* Sydney, George Allen and Unwin

Buxton, R, G Brennan, J Engleke, E Jack and M Stafford Smith (1995) *DroughtPlan Regional Report: Kimberley* Alice Springs, CSIRO

Campbell, R, P Crowley and P Demura (1983) "Impact of drought on national income and employment" *Quarterly Review of the Rural Economy* 5(3)

Carter, J, W Hall, K Brook, G McKeon, K Day and C Paull (2000) "Aussie GRASS: Australian grassland and rangeland assessment by spatial simulation" in G L Hammer, N Nicholls and C Mitchell (Eds), *Applications of seasonal climate forecasting in agricultural and natural ecosystems* Dordrecht, Kluwer Academic Publishers: 329-349

Cate, F H (1996) "Communications, policy-making, and humanitarian crises" in R I Rotberg and T G Weiss (Eds), *From Massacres to Geneocide. The Media, Public Policy, and Humanitarian Crises* Washington, The Brookings Institution

Chapman, T G (Ed.) (1976) *Drought* Canberra, Australian Government Publishing Service

Chiew, F H S, T A Piechota, J A Dracup and T A McMahon (1998) "El Niño/Southern Oscillation and Australian rainfall, stream flow and drought: Links and potential for forecasting" *Journal of Hydrology* 204: 138-149

Christenson, J A and C B Flora (1991) "A rural policy agenda for the 1990s" in C B Flora and J A Christenson (Eds), *Rural Policies for the 1990s* Boulder, CO, Westview Press: 333-337

Clark, A, T Brinkley, B Lamont and G Laughlin (2000) *Exceptional Circumstances: A case study in the application of climate information to decision making* Proceedings of the Cli-Manage Conference Albury, NSW

Clark, A J and T Brinkley (2001) *Risk Management for Climate Agriculture and Policy* Canberra, Bureau of Rural Sciences, Commonwealth of Australia

Clewett, J F, N M Clarkson, D A George, S H Ooi, D T Owens, I J Partridge and G B Simpson (2003) *Rainman StreamFlow version 4.3: A comprehensive climate and stream flow analysis package on CD to assess seasonal forecasts and manage climatic risk* QI03040 Queensland, Department of Primary Industries

Coakes, S and M Fisher (2001) *Risk perception, farmers and biotechnology* Bureau of Rural Sciences Report prepared for AFFA Canberra

Cohen, K (2002) "Letter to editor" *Sydney Morning Herald* 12th July 2002

Colebatch, H (2002) *Policy* (Second Edition) Buckingham, Open University Press

Collins, S t H B (1995) *Response to Question without Notice by Minister for Primary Industries and Energy* Senate Hansard 30 May 1995

Colls, K and R Whitaker (1990) *The Australian Weather Book* Brookvale, National Book Distributors

Commonwealth Department of Human Services and Health (1994) *Working towards a regional focus* Report on the regional focus evaluation Canberra, Australian Government Publishing Service, January

Conacher, A and J Conacher (1995) *Rural Land Degradation in Australia*, Oxford University Press, Australia

Cox, P G (1996) "Some issues in the design of agricultural decision support systems" *Agricultural Systems* 52(2/3): 355-381

Cox, P G, A D Shulman, P E Ridge, M A Foale and A L Garside (1995) "An interrogative approach to system diagnosis: An invitation to the dance" *Journal for Farming Systems Research-Extension* 5: 67-83

Crean, S, the Hon MP (1992) *Rural Adjustment Bill 1992: Second Reading Speech* House of Representatives Hansard 3 November 1992

CSIRO (2001) *Climate change projections for Australia* CSIRO Atmospheric Research

Daiyi, N, L Ford and D Rose (2002) "Life in Country: Ecological Restoration on Aboriginal Land" *Cultural Survival Quarterly* 26(2): 37-38

Daly, D (1994) *Wet as a shag. Dry as a bone. Drought in a variable climate* QDPI Information series QI93028 Brisbane, Queensland Department of Primary Industries

Damasio, A R (1994) *Descartes' error: Emotion, reason and the human brain* New York, Avon

Davies, S (2000) "Effective Drought Mitigation: Linking Micro and Macro Levels" in D A Wilhite (Ed.) *Drought: A Global Assessment* London, Routledge Publishers Volume 2: 3-16

de Jager, J M, A B Potgieter and W J van den Berg (1998) "Framework for forecasting the extent and severity of drought in maize in the Free State province of South Africa" *Agricultural Systems* 57(351-365)

Delbridge, A, J R L Bernard, D Blair, P Peters and S Butler (1991) *The Macquarie Dictionary* (Second Edition), The Macquarie Library Pty Ltd.

Dinar, A and A Keck (2000) "Water Supply Variability and Drought Impact and Mitigation in Sub-Saharan Africa" in D A Wilhite (Ed.) *Drought: A Global Assessment* London, Routledge Publishers Volume 2: 129-148

Dodson, J R and M Westoby (1985) "Are Australian Ecosystems Different?" *Proceedings of the Ecological Society of Australia* 14

Dolowitz, D P and D Marsh (2000) "Learning from Abroad: The Role of Policy Transfer in Contemporary Policy-Making" *Governance* 13(1): 5-24

Donald, C M (1982) "Innovation in Australian Agriculture" in D B Williams (Ed.) *Agriculture in the Australian Economy* Sydney, Sydney Univerity Press: 55-82

Donnelly, J R, M Freer and A D Moore (1997) "GRAZPLAN: Decision support systems for Australian grazing enterprises. I. Overview of the GRAZPLAN project and a description of the MetAccess and LambAlive DSS" *Agricultural Systems* 54: 57-76

Douglas, M and A Wildavsky (1982) *Risk and Culture: An Essay on the Selection of Technical and Environmental Dangers* Berkeley, University of California Press

Douglas, R, M Oliver and N Hall (2002) *Farm Management Deposits Scheme Evaluation*, Rural Economic Services

Downs, A (1972) "Up and down with ecology: The issue attention cycle" *Public Interest* 28(1): 38-50

DPIE (Department of Primary Industries and Energy) (1996) *Mid Term Review of the Rural Adjustment Scheme: Submission by the Department of Primary Industries and Energy* Canberra

DPRTF (Drought Policy Review Task Force) (1990) *National Drought Policy* Canberra, AGPS

Dracup, J, K S Lee and E G Paulson (1980) "On the Definition of Droughts" *Water Resources Research* 16(2): 297-302

Drought Policy Task Force (1997) *Review of the National Drought Policy* Task Force of Officials from the Commonwealth, State and Territory Governments Canberra

du Pisani, L G, H J Fouché and J C Venter (1998) "Assessing rangeland drought in South Africa" *Agricultural Systems* 57: 367-380

Edelman, M (1988) *Constructing the Political Spectacle* Chicago, University of Chicago Press

Elliott, D (1989) "Tales from the darkside: Ethical implications of disaster coverage" in L M Waters, L Weilkins and T Walters (Eds), *Bad Tidings. Communication and Catastrophe* Hillsdale, Lawrence Erlbaum

Ellis, F (2000) *Rural livelihoods and diversity in developing countries* Oxford, Oxford University Press

Ericson, R, P Baranek and J B L Chan (1987) *Visualising Deviance: A Study of News Organisation* Milton Keynes, Open University Press

Ernst & Young (2000) *Multi Peril Crop Insurance Project: Phase 2 Report* Agriculture, Fisheries & Forestry

European Commission (1999) *Income Insurance in European Agriculture*

Ewart, J (2002) *Prudence not prurience: A framework for journalists reporting disasters* Paper read to the annual conference of the Australian and New Zealand Communication Association Bond University http://www.bond.edu.au/hss/communication/ANZCA/papers/JEwartPaper.pdf

Federal Emergency Management Agency (1995) *National Mitigation Strategy* Washington, DC, 1995

Finucane, M (2000) *Improving Quarantine Risk Communication: Understanding Public Risk Perceptions* Decision Research Report 00-7 Eugene, Oregon

Finucane, M L, E Peters and P Slovic (2003) "Judgement and decision making: the dance of affect and reason" in S Schneider and J Shanteau (Eds), *Emerging perspectives on judgement and decision research* New York, Cambridge University Press: 327-364

Fitzpatrick, E A and H A Nix (1970) "The climatic factor in Australian ecology" in R Milton Moore (Ed.) *Australian Grasslands* Canberra, Australian National University Press: 3-26

Flannery, T F (1994) *The future eaters, an ecological history of the Australasian lands and people* Sydney, Reed Books

Flinn, W L and D E Johnson (1974) "Agrarianism Among Wisconsin Farmers" *Rural Sociology* 39(2): 187-204

Foley, J C (1957) *Droughts in Australia. Review of records from earliest settlement to 1955* Bulletin No 43 Melbourne, Bureau of Meteorology

Fouché, H J, J M de Jager and D P J Opperman (1985) "A mathematical model for assessing the influence of stocking rate on the incidence of droughts and for estimating the optimal stocking rates" *Journal of the Grassland Society of Southern Africa* 2(3): 3-6

Freebairn, J (2002a) *Drought* Connections, Newsletter of Agribusiness Association of Australia and the Australian Agricultural and Resource Economics Society. Spring (December): 3

Freebairn, J (2002b) "Drought" *Connections, Newsletter of Agribusiness Association of Australia and the Australian Agricultural and Resource Economics Society*

Freebairn, J (2003) "Economic policy for rural and regional Australia" *Australian Journal of Agricultural and Resource Economics* 47(3): 389-414

Freebairn, J W (1978) "Pros and Cons of Temporary Industry Assistance" *Australian Journal of Agricultural Economics* 22(3): 194-205

Freebairn, J W (1983) "Drought assistance policy" *Australian Journal of Agricultural Economics* 27(3): 185-199

Friedel, M H, B D Foran and D M Stafford Smith (1990) "Where the creeks run dry or ten feet high: Pastoral management in arid Australia" *Proceedings of the Ecological Society of Australia* 16: 185-194

Gardner, B L (1994) "Crop Insurance in U.S. Farm Policy" in D L Heuth and W H Furtan (Eds), *Economics of Agricultural Crop Insurance: Theory and Evidence* Boston/Dordrecht/London, Kluwer Academic Publishers: 18-44

Georgia Department of Natural Resources (2003) *Georgia Drought Management Plan* Atlanta

Gibbs, W J and J V Maher (1957) *Droughts in Australia: Bulletin No 43* Commonwealth Bureau of Meteorology Melbourne

Gigerenzer, G and P M Todd (1999) *Simple Heuristics That Make Us Smart*, Oxford University Press

Ginytjirrang Mala (1994) *An Indigenous Marine Protection Strategy for Manbuynga ga Rulyapa* with the assistance of A.D.V.Y.Z for the Northern Land Council and Ocean Rescue 2000 - November 1994, unpublished manuscript

Gitlin, T (1980) *The Whole World is Watching* Berkeley, University of California Press

Gittins, R (2002) "Farmers who fail don't deserve pity" *Sydney Morning Herald* 16 October 2002

Glantz, M H (1987) *Drought and Hunger in Africa: Denying Famine a Future* Cambridge, Cambridge University Press

Glantz, M H (2000) "Drought follows the plough: A cautionary note" in D A Wilhite (Ed.) *Drought: A Global Assessment* London, Routledge Volume 2: 285-291

Glantz, M, R Katz and M Krenz (Eds) (1987) *The Societal Impacts Associated with the 1982-83 Worldwide Climate Anomalies* Boulder, Colorado, National Center for Atmospheric Research

Glantz, M H, R W Katz and N Nicholls (Eds) (1991) *Teleconnections Linking Worldwide Climate Anomalies* Cambridge, Cambridge University Press

Goltz, J D (1984) "Are the news media responsible for disaster myths? A content analysis of emergency response imagery" *International Journal of Mass Emergencies and Disasters* 2(3): 345-368

Goodwin, B K and V H Smith (1995) *The Economics of Crop Insurance and Disaster Aid* Washington D C, The AEI Press

Gould, R (1982) "To Have and Have Not: The Ecology of Sharing Among Hunter-Gatherers" in N Williams and E Hunn (Eds), *Resource Managers: North American and Australian Hunter-Gatherers* Canberra, Aboriginal Studies Press: 69-112

Gow, J (1994-95) "Drought, lies and videotape" *Policy and Politics* 10(4): 7-11

Gow, J (1997) "Commonwealth Drought Policy: 1989-1995. A Case Study of Economic Rationalism" *Australian Journal of Social Issues* 32(3): 272-282

Graber, D (2001) *Processing Politics* Chicago, University of Chicago Press

Gray, I (1991) *Politics in Place: Social Power Relations in an Australian Country Town* Cambridge, Cambridge University Press

Gray, I and E Phillips (2001) "Beyond life in 'the bush': Australian rural cultures" in S Lockie and L Bourke (Eds), *Rurality Bites: The Social and Environmental Transformation of Rural Australia* Annandale, NSW, Pluto Press: 52-59

Greaves, T (1998) "Water Rights in the Pacific Northwest" in J Donohue and B Johnston (Eds), *Water, Culture and Power: Local Struggles in a Global Context* Washington DC, Island Press: 35-46

Groves, R (1994) *Australian Vegetation* Cambridge, Cambridge University Press

Gruen, D and S Shretha (Eds) (2000) *The Australian Economy in the 1990s* Sydney, Reserve Bank of Australia

Gudger, M (1991) *Crop Insurance: Failure of the Public Sector and the Rise of the Private Sector Alternative* Risk in Agriculture. Proceedings of the Tenth Agriculture Sector Symposium Washington, The World Bank

Guttman, N B (1999) "Accepting the Standardized Precipitation Index: A calculation algorithm" *Journal of the American Water Resources Association* 35(2): 311-322

Haberkorn, G, G Hugo, M Fisher and R Aylward (1999) *Country Matters: Social Atlas of Rural and Regional Australia* Canberra, Bureau of Rural Sciences

Hall, W, D Bruget, J Carter, G McKeon, J Yee Yet, A Peacock, R Hassett and K Brook (2001) *Australian Grassland and Rangeland Assessment by Spatial Simulation (Aussie GRASS)* QNR9 Final Report for the Climate Variability in Agriculture Program Queensland, Department of Natural Resources and Mines

Hamblin, A and G Kyneur (1993) *Trends in wheat yields and soil fertility in Australia* Canberra, Australian Government Printing Service

Hamilton, C (1996) *Economic Rationalism in the Bush* Paper presented to the Commonwealth Department of Primary Industries and Energy Canberra, 19 November

Hammer, G L, D P Holzworth and R Stone (1996) "The value of skill in seasonal climate forecasting to wheat crop management in a region with high climate variability" *Australian Journal of Agricultural Research* 47: 717-737

Hammer, G L, N Nicholls and C Mitchell (2000) *Applications of Seasonal Climate Forecasting in Agricultural and Natural Ecosystems—The Australian Experience* Atmospheric and Oceanographic Sciences Library, Volume 21 Dordrecht, Kluwer Academic Publishers

Hammond, K R (1996) *Human Judgement and Social Policy. Irreducible Uncertainty, Inevitable Error, Unavoidable Injustice* New York, Oxford University Press

Hancock, L (2002) "The care crunch: Changing work, families and welfare in Australia" *Critical Social Policy* 22(1): 119-140

Hardaker, J B, R B M Huirne and J R Anderson (1997) *Coping with Risk in Agriculture* Wallingford, CAB International

Hattis, T R (1994) "Communicating environmental risk" *Risk Analysis* 5: 704-760

Hayman, P T (2001) *Farmers and Agricultural Scientists in a Variable Climate* Unpublished PhD Thesis University of Western Sydney

Hazell, P B R (1992) "The Appropriate Role of Agricultural Insurance in Developing Countries" *Journal of International Development* 4: 567-581

Heathcote, L (2002) *Braving the Bull of Heaven* Presentation to Royal Geographical Society of Queensland http://www.rgsq.gil.com./heath2c.htm

Heathcote, R L (1973) "Drought perception" in J V Lovett (Ed.) *The environmental, economic and social significance of drought* Sydney, Angus & Robertson: 17-40

Heathcote, R L (2000) "'She'll be right, mate': Coping with Drought—Strategies old and new in Australia" in D A Wilhite (Ed.) *Drought: A Global Assessment* London, Routledge Volume 2: 59-69

Hennessy, K J, R Suppiah and C M Page (1999) "Australian rainfall changes, 1910-1995" *Australian Meteorological Magazine* 48: 1-14

Hercus, L and P Clark (1986) "Nine Simpson Desert wells" *Archaeology in Oceania* 21: 51-62

Higgins, V (2001) *Smoothing the Process of Change? A Genealogy of Farm Viability in Australia (1967-1997)* Unpublished PhD thesis Central Queensland University

Higgins, V and D Stehlik (1999) *Ageing at the Margins? Self-Reliance and Rural Restructuring as Agendas for Aged Care* Paper presented at the National Social Policy Conference—Social Policy for the 21st Century. Justice and Responsibility University of New South Wales, Sydney

Hobbs, J E (1998) "Present climates of Australia and New Zealand" in J E Hobbs, J A Lindesay and H Bridgman (Eds), *Climates of the Southern Continents: Present, Past and Future* Chichester, John Wiley & Sons: 18-23

Hobbs, J E, J A Lindesay and H A Bridgman (Eds) (1998) *Climates of the Southern Continents: Present, Past and Future* Chichester, John Wiley & Sons

Holmes, J H (1997) "Diversity and change in Australia's rangeland regions: Translating resource values into regional benefits" *The Rangeland Journal* 19: 3-25

Houghton, J T (1994) *Global Warming: The Complete Briefing* Oxford, Lion Publishing

Houghton, J T, Y Ding, D J Griggs, M Noguer, P J van der Linden and D Xiaosu (Eds) (2001) *Climate Change 2001: The Scientific Basis* Cambridge, Cambridge University Press

Howlett, M and M Ramesh (1995) *Studying Public Policy. Policy Cycles and Policy Subsystems* Toronto, Oxford University Press

Hunn, E (1993) "What is Traditional Ecological Knowledge?" in N Williams and G Baines (Eds), *Traditional Ecological Knowledge: Wisdom for Sustainable Development*, Centre for Resource and Environmental Studies, The Australian National University: 13-15

Hy, R J and W L Waugh, Jr (1990) "The Function of Emergency Management" in W L Waugh, Jr and R J Hy (Eds), *Handbook of Emergency Management: Programs and Policies Dealing with Major Hazards and Disasters* New York, Greenwood Press: 11-26

Industries Assistance Commission (IAC) (1983) *Rural Adjustment. (Interim Report)* Canberra, Australian Government Publishing Service

Industries Assistance Commission (1996) *Crop and Rainfall Insurance* Canberra, Australian Government Publishing Service

Industry Commission (1995) *Assistance to agricultural and manufacturing industries* Canberra, Australian Government Publishing Service

Industry Commission (1996) *Submission to the Mid-Term Review of the Rural Adjustment Scheme* Canberra, Commonwealth of Australia

Industry Commission (1998) *Microeconomic Reforms in Australia: A Compendium from the 1970s to 1997 Research Paper* Canberra, Australian Government Publishing Service

International Panel on Climate Change (IPCC) (2001) *Climate Change 2001: Synthesis Report* Cambridge, Cambridge University Press

ISDR Drought Discussion Group (2003) *Drought. Living with Risk: An Integrated Approach to Reducing Societal Vulnerability to Drought* Geneva, International Strategy for Disaster Reduction

Jamrozik, A (1983) "Universality and Selectivity. Social Welfare in a Market Economy" in A Graycar (Ed.) *Retreat from the Welfare State* Sydney, University of New South Wales Press: 171-188

Jeffrey, S J, J O Carter, K B Moodie and A R Beswick (2001) "Using spatial interpolation to construct a comprehensive archive of Australian climate data" *Environmental Modelling & Software* 16: 309-330

Jemphrey, A, A Berrington and D a t p . (2000) "Surviving the media: Hillsborough, Dunblane and the press" *Journalism Studies* 1(3): 469-483

Jennings, G and D Stehlik (2000) "Agricultural women in Central Queensland and changing modes of production: a preliminary exploration of the issues" *Rural Society* 10(1): 63-78

John, P (1998) *Analysing Public Policy* London, Pinter

Jones, R and B Meehan (1997) "Balmarrk wana: Big winds of Arnhem Land" in E K Webb (Ed.) *Windows on Meteorology: Australian Perspective* Melbourne, CSIRO Publishing: 14-19

Karoly, D, J Risbey and A Reynolds (2003) *Global warming contributes to Australia's worst drought* Sydney, WWF Australia http://www.wwf.org.au

Kay, J (2003) *The Truth About Markets. Their Genius, their Limits, their Follies* London, Allen Lane

Keating, B A and H Meinke (1998) "Assessing Exceptional Drought with a cropping systems simulator: A case study for grain production in north-east Australia" *Agricultural Systems* 57: 315-332
Ker Conway, J (1989) *The road from Coorain— an Australian memoir* London, Heinemann

Kerin, J (1987) *The Rural Book. A Guide for Rural People about Major Commonwealth Services & Programs* Canberra, Australian Government Publishing Service

Kimber, R (1986) *Man from Arltunga: Walter Smith, Australian Bushman* Carlisle, Hesperian Press

Kimber, R (1997) "Cry of the Plover, Song of the desert rain" in E K Webb (Ed.) *Windows on Meteorology: Australian Perspective* Melbourne, CSIRO Publishing: 7-13

King, R (1994) "Suicide Prevention: Dilemmas and some solutions" *Rural Society* 4(3/4): 2-6

Kingsford, R, A Boulton and J Puckridge (1998) "Challenges in managing dryland rivers crossing political boundaries: Lessons from Cooper Creek and the Paroo River, central Australia" *Aquatic Conservation: Marine and Freshwater Ecosystems* 8: 361-378

Knight, F H (1921) *Risk, Uncertainty and Profit* New York, Century Press

Knutson, C, M Hayes and T Phillips (1998) *How to Reduce Drought Risk* A guide prepared by the Preparedness and Mitigation Working Group of the Western Drought Coordination Council Lincoln, Nebraska, National Drought Mitigation Center

Koch, T (1989) "Harper ignored DPI for John Relative" *Courier-Mail* 17 April 1989

Kraft, D and R Piggott (1989) "Why Single Out Drought?" *Search* 20(6): 189-192

Langton, M (2002) *Freshwater* Background briefing papers Broome, WA, Lingiari Foundation
http://www.atsic.gov.au/issues/Indigenous_Rights/Indigenous_Rights_Waters/docs/layout_papers.pdf

Latz, P (1995) *Fire in the desert: Increasing biodiversity in the short term, decreasing it in the long term* Country in Flames; Proceedings of the 1994 symposium on biodiversity and fire in North Australia Canberra and Darwin, Biodiversity Unit, Department of the Environment, Sport and Territories, and the North Australia Research Unit

Laughlin, G P, H Zuo, J Walcott and A Bugg (2003) "The Rainfall Reliability Wizard—A new tool to rapidly analyse spatial rainfall reliability with examples" *Environmental Modelling and Software* 18: 49-57

Lawson, H (1988) "Beaten Back" in D Baglin (Ed.) *Henry Lawson's Images of Australia* Frenchs Forest, NSW, Reed Books

Lewis, H (1993) "Traditional Ecological Knowledge - Some Definitions" in N Williams and G Baines (Eds), *Traditional Ecological Knowledge: Wisdom for Sustainable Development* Canberra, Centre for Resource and Environmental Studies, The Australian National University: 8-12

Linacre, E T and B Geerts (1997) *Climates and Weather Explained* London, Routledge

Lindblom, C E (1959) "The Science of "Muddling Through" *Public Administration Review* 19: 79-82

Lindblom, C E (2002) *The Market System. What It Is, How It Works, and What To Make of It* New Haven, Yale University Press

Lindesay, J A and K M Johnson (2003) *Changing seasonality in Australian rainfall* Preprints of the 7th International Conference on Southern Hemisphere Meteorology and Oceanography Wellington, New Zealand, American Meteorological Society, Boston

Lingiari Foundation (2002) *Onshore: Water rights discussion booklet* Broome, WA, Lingiari Foundation http://www.atsic.gov.au/issues/Indigenous_Rights/Indigenous_Rights_Waters/docs/layout_papers.pdf

Lloyd, B, MP (1992) *Rural Adjustment Bill 1992: Second Reading Debate* House of Representatives Hansard Canberra 10 November 1992

Lockie, S (2000) "Crisis and conflict: Shifting discourses of rural and regional Australia" in B Pritchard and P McManus (Eds), *Land of discontent: The dynamics of change in rural and regional Australia* Sydney, UNSW Press: 14-32

Lovett, J V (Ed.) (1973) *The environmental, economic and social significance of drought* Sydney, Angus & Robertson

Lowe, P (1990) *Jilji: Life in the Great Sandy Desert* with Jimmy Pike Broome, Magabala Books

Malcolm, L R (1992) *Farm risk management and decision making* Proceedings of national workshop on risk management, 9-11 November 1992 Melbourne, Department of Natural Resources and Environment

Malcolm, L R (1994) *Managing farm risk: There may be less to it than is made of it* Proceedings of Conference: Risk Management in Australian Agriculture The University of New England, Armidale, NSW

Mangkaja Arts Resource Agency (2003) *Martuwarra and Jila, River and Desert* Perth, Mungkaja Arts Resource Agency Aboriginal Corporation and the University of Western Australia

March, J G and H A Simon (1958) *Organizations* Carnegie Institute of Technology, Pittsburgh. Graduate School of Industrial Administration. Publications New York, J. Wiley

Marsh, S P and D J Pannell (2000) "Agricultural extension policy in Australia: The good, the bad and the misguided" *The Australian Journal of Agricultural and Resource Economics* 44(4): 605-627

Marston, G (2000) "Metaphor, morality and myth: A critical discourse analysis of public housing policy in Queensland" *Critical Social Policy* 20(3): 349-373

Martin, P (1995) "Drought. Impact on farm financial performance" in Australian Bureau of Agricultural and Resource Economics (Ed.) *Farm Surveys Report* Canberra: 58-64

Martin, P, S Hooper, A Blias, N Hanna and M Ford (2003) "Farm Financial Performance" in ABARE Economics (Ed.) *Australian Farm Surveys Report* Canberra: 1-14

Matthews, R A J (2000) "Facts versus factions: The use and abuse of subjectivity in scientific research" in J Morris (Ed.) *Rethinking risk and the precautionary principle* Oxford, Butterworth-Heinemann: 247-282

Mauldon, R G and H P Schapper (1974) *Australian farmers under stress in prosperity and recession* Nedlands, University of Western Australia Press

May, P J (1992) "Policy Learning and Failure" *Journal of Public Policy* 12(4): 331-354

Mayer, H (1994) "Australian mass media and natural disasters" in R Tiffen (Ed.) *Mayer on the Media. Issues and arguments* St Leonards, Allen and Unwin: 141-151

Mayers, B (1995) *Insurance Based Risk Management for Drought* Occasional Paper CV02/95 Canberra, Land and Water Resources Research and Development Corporation and Rural Industries Research and Development Corporation

Mayers, B (1996) *Of Droughts and Flooding Rains* Land and Water Resources Research and Development Corporation and Rural Industries Research and Development Corporation Canberra

McCall, M W and R E Kaplan (1990) *Whatever it takes—The realities of managerial decision making* New Jersey, Prentice-Hall

McColl, J C, R Donald and C Shearer (1997) *Rural Adjustment: Managing Change* Commonwealth of Australia Canberra

McCombs, M and A Reyolds (2002) "News influence on our pictures of the world" in J Bryant and D Zillman (Eds), *Media Effects* Mahwah, NJ, Erlbaum

McKenzie, D (1996) "Welfare to replace farm drought aid" *The Weekend Australian* September 28-29

McKeon, G M, S M Howden, N O J Abel and J M King (1993) *Climate change: Adapting tropical and subtropical grasslands* Proceedings of the XVIIth International Grassland Congress Palmerston North, New Zealand

McMichael, P (1984) *Settlers and the agrarian question: Capitalism in colonial Australia* Cambridge, Cambridge University Press

McVicar, T R and D L B Jupp (1998) "The current and potential operational uses of remote sensing to aid decision on Drought Exceptional Circumstances in Australia: A review" *Agricultural Systems* 57: 399-468

Meadows, M (2001) "A return to practice: Reclaiming journalism as a public conversation" in S Tapsall and C Varley (Eds), *Journalism Theory in Practice* Melbourne, Oxford University Press

Megalogenis, G and A Wahlquist (2002) "Our Greenest Drought" *The Weekend Australian* October 26-27

Mill, J S (1893) *Principles of Political Economy* New York, D Appleton and Company

Miranda, M J and J W Glauber (1997) "Systemic Risk, Reinsurance and the Failure of Crop Insurance Markets" *American Journal of Agricultural Economics* 79: 206-215

Monnik, K (2000) "Role of Drought Early Warning Systems in South Africa's Evolving Drought Policy" in D A Wilhite, M V K Sivakumar and D A Wood (Eds), *Early Warning Systems for Drought Preparedness and Drought Management, Proceedings of an Expert Group Meeting, Lisbon, Portugal, 5-7 September* Geneva, Switzerland, World Meteorological Organization: 47-56

Montmarquet, J A (1989) *The Idea of Agrarianism* Moscow, Idaho, University of Idaho Press

Moss, D A (2002) *When all else fails: Government as the ultimate risk manager* Cambridge USA, Harvard University Press

Moyer, H W and T E Josling (1990) *Agricultural Policy Reform: Politics and Process in the EC and the USA* New York, Harvester Wheatsheaf

Mules, W, T Schirato and B Wigman (1995) "Rural identity within the symbolic order: media representations of the drought" in P Share (Ed.) *Communication and Culture in Rural Areas* Wagga Wagga, Charles Sturt University Centre for Rural Social Research

Mullen, J D, D Vernon and K I Fishpool (2000) "Agricultural extension policy in Australia: Public funding and market failure" *Australian Journal of Agricultural and Resource Economics* 44(4): 629-645

Multi Peril Crop Insurance Task Force (2003) *Final Report*, Agriculture, Forestry and Fisheries, Western Australia

Munro, R K and M J Lembit (1997) *Managing climate variability in the national interest: Needs and objectives* Climate prediction for agricultural and resource management: Australian Academy of Science Conference Canberra

Murrell, B (1984) "Hydrological Regimes in the Australia Arid Zone with Notes on the Current changes in Hydrological Regime in the Willonson Ranges, South Australia" in H C Cogger and E E Cameron (Eds), *Arid Australia* Sydney, Australian Museum: 327-334

Musgrave, W, P Rickards and I Whan (1975) "Poverty among farmers in New South Wales and Queensland" in R F Henderson (Ed.) *Financial aspects of rural poverty* Canberra, AGPS

Myers, F (1982) "Always Ask: Resource use and land ownership among the Pintupi Aborigines of the Australian Western Desert" in N Williams and E Hunn (Eds), *Resource Managers: North American and Australian Hunter-Gatherers* Canberra, Aboriginal Studies Press: 173-195

National Drought Policy Commission (2000) *Preparing for Drought in the 21st Century: Executive Summary* Washington, D C, 2000

National Farmers Federation (2002) "The ongoing battle to overhaul drought policy. An NFF discussion paper" *Reform* Autumn: 7-10

Natsios, A (1996) "Illusions of influence. The CNN effect in complex emergencies" in R I Rotberg and T G Weiss (Eds), *From Massacres to Genocide* Washington, Brookings Institution

Nelson, R A, D P Holzworth, G L Hammer and P T Hayman (2002) "Infusing the use of seasonal climate forecasting into crop management practice in North East Australia using discussion support software" *Agricultural Systems* 74: 393-414

Nicholls, N (1997a) "The centennial drought" in E K Webb (Ed.) *Windows on Meteorology—Australian perspective* Collingwood, CSIRO: 183-204

Nicholls, N (1997b) "Developments in climatology in Australia: 1946-1996" *Australian Meteorological Magazine* 46: 127-135

Nicholls, N (1999) "Cognitive illusions, heuristics and climate prediction" *Bulletin of the American Meteorological Society* 80: 1385-97

Nicholls, N and A H Sellers (1991) "The El Niño/Southern Oscillation and Australian vegetation" *Advances in Vegetation Science* 12: 23-36

Nicholls, N and K K Wong (1990) "Dependence of rainfall variability on mean rainfall, latitude, and the Southern Oscillation" *Journal of Climate* 3: 163-170

National Land and Water Resources Audit (NLWRA) (2002) *Australia's Natural Resources: 1997-2002 and Beyond* Canberra, NLWRA

NSW Farmers' Association (2001) *Submission to the review of natural disaster relief and mitigation arrangements*

NSW Farmers' Association (2003) *Preparing for Drought: The Key to Sustainability* Discussion Paper September 2003 Sydney

O'Meagher, B (2003) "Economic aspects of drought and drought policy" in L C Botterill and M Fisher (Eds), *Beyond Drought: People, Policy and Perspectives* Melbourne, CSIRO Publishing: 109-130

O'Meagher, B, L G du Pisani and D H White (1998) "Evolution of Drought Policy and Related Science in Australia and South Africa" *Agricultural Systems* 57(3): 231-258

O'Meagher, B, M Stafford Smith and D H White (2000) "Approaches to Integrated Drought Risk Management: Australia's National Drought Policy" in D A Wilhite (Ed.) *Drought: A Global Assessment* London, Routledge Volume 2: 115-128

O'Neil, M, A Kosturjak, J Molloy, S Lindsay, M Bright and J Weatherford (2000) *Mid-term Review of Farm Family Restart Scheme* South Australian Centre for Economic Studies Adelaide

Organisation for Economic Cooperation and Development (2003) *Agricultural Policies in OECD Countries. Monitoring and evaluation 2003. Highlights* Paris, OECD

Pannell, D J, L R Malcolm and R S Kingwell (2000) "Are we risking too much? Perspectives on risk in farm modelling" *Agricultural Economics* 23(1): 69-78

Parham, D (2002) *Microeconomic reforms and the revival in Australia's growth in productivity and living standards* Paper presented to the Conference of Economists Adelaide

Parker, K (1905) *The Euahlayi Tribe: A Study of Aboriginal Life in Australia* London, Archibald Constable & Co

Parsons, W (1995) *Public Policy. An introduction to the theory and practice of policy analysis* Cheltenham, Edward Elgar

Patton, D (1993) "The ABCs of risk assessment" *EPA Journal; Environmental Protection Agency, Washington, DC* Jan-March: 10-15

Peart, G (1992) *Pastures, livestock and the bottom line* Proceedings of the 7th annual conference of the grasslands conference Tamworth

Peterson, N (1976) "The natural and cultural areas of Aboriginal Australia: A preliminary analysis of populations groups with adaptive significance" in N Peterson (Ed.) *Tribes and Boundaries in Australia* Canberra, Australian Institute of Aboriginal Studies: 50-71

Pinker, R (1973) *Social Theory and Social Policy* London, Heinemann

Power, S B, T Casey, C Folland, A Colman and V Mehta (1999) "Inter-decadal modulation of the impact of ENSO on Australia" *Climate Dynamics* 15: 319-324

Power, S B, F Tseitkin, V Mehta, B Lavery, S Torok and N Holbrook (1999) "Decadal climate variability in Australia during the twentieth century" *International Journal of Climatology* 19: 169-184

Pritchard, B (2002) *"Of droughts and flooding rains"; Policy on Rural Australia* The Drawing Board: Policy on Rural Australia, University of Sydney
http://www.econ.usyd.edu.au/drawingboard/digest/0212/pritchard.htm

Productivity Commission (1998) *Review of NSW Rural Assistance Act 1989* Canberra, Australian Government Publishing Service

Purtill, A, M Backhouse, A Abey and S Davenport (1983) "A study of the drought" *Quarterly Review of the Rural Economy* 5(1): 3-11

Pusey, M (1991) *Economic rationalism in Canberra: A nation-building state changes its mind* New York, Cambridge University Press

Pusey, M (2003) *The Experience of Middle Australia. The Dark Side of Economic Reform* Cambridge, Cambridge University Press

Putnam, L P (2002) "Framing environmental conflicts: The Edwards Aquifer dispute" in E Gilboa (Ed.) *Media and Conflict. Framing issues. Making policy. Shaping opinions* Ardlsey Park, Transnational Publishers

Quarantelli, E L (1989) "The social science study of disasters and mass communication" in L M Waters, L Weilkins and T Walters (Eds), *Bad Tidings. Communication and Catastrophe* Hillsdale, Lawrence Erlbaum

Queensland Government (1992) *Drought—Managing for Self Reliance. A Policy Paper* Brisbane, March

Quiggin, J (1986) "A Note on the Variability of Rainfall Insurance" *Australian Journal of Agricultural Economics* 30: 63-69

Quiggin, J (1994) "The Optimal Design of Crop Insurance" in D L Heuth and W H Furtan (Eds), *Economics of Agricultural Crop Insurance: Theory and Evidence* Boston/Dordrecht/London, Kluwer Academic Publishers: 115-134

Rabat Declaration (2001) *Meeting on Opportunities for Sustainable Investment in Rainfed Areas of West Asia and North Africa* Rabat, Morocco

Radcliffe-Brown, A R (1930) "The rainbow-serpent myth in south-east Australia" *Oceania* 1(3): 342-347

Rein, M and D Schön (1993) "Reframing policy discourse" in F Fischer and J Forester (Eds), *The Argumentative Turn in Policy Analysis and Planning* Durham, Duke University Press

Ribot, J C (1996) "Introduction. Climate variability, climate change and vulnerability: Moving forward by looking back" in J C Ribot, A R Magalhães and S S Panagides (Eds), *Climate variability, climate change and social vulnerability in the semi-arid tropics* Cambridge, Cambridge University Press

Robbins, P F, N Abel, H Jiang, M Mortimer, M Mulligan, G S Okin, D M Stafford Smith and B L Turner, II (2002) "Group 2: What Are the Key Dimensions at the Community Scale?" in J F Reynolds and D M Stafford Smith (Eds), *Dahlem Workshop Report 88.* Berlin, Dahlem University Press: 325-355

Robinson, P (2002) "Global television and conflict resolution: Defining the limits of the CNN effect" in E Gilboa (Ed.) *Media and Conflict. Framing issues. Making policy. Shaping opinions* Ardsley Park, Transnational Publishers

Rodgers, S (1996) "Two Australias" *Courier-Mail* March 23

Rogers, E M and J W Dearing (1988) "Agenda-setting research: Where has it been? Where is it going?" in J Anderson (Ed.) *Communication Yearbook* Beverly Hills, Sage

Rose, D (1999) "Indigenous Ecologies and an Ethic of Connection" in N Low (Ed.) *Global Ethics for the 21st Century* London, Routledge: 175-86

Rose, D (2000) *Dingo Makes Us Human: Life and Land in an Australian Aboriginal Culture* (Paperback edition) Cambridge, Cambridge University Press

Rose, D (2004 in press) "Rhythms, Patterns, Connectivities: Indigenous Concepts of Seasons and Change, Victoria River district, NT" in T Sherratt, T Griffiths and L Robin (Eds), *A change in the weather: Climate and culture in Australia* Sydney, Halstead Press

Rose, D, D James and C Watson (2003) *Indigenous Kinship with the Natural World* Sydney, National Parks and Wildlife Service, NSW

Rose, R (1991) "What is Lesson-Drawing?" *Journal of Public Policy* 11(1): 3-30

Rose, R (1993) *Lesson-drawing in Public Policy: A Guide to Learning across Time and Space* Chatham, New Jersey, Chatham House Publishers

Rowlands, D (2000/01) "Purchaser-provider in social policy delivery: How can we evaluate the Centrelink arrangements?" *Australian Social Policy*: 69-87

Rural Adjustment Act 1992 No. 240 of 1992

Saji, N H, B M Goswami, P N Vinayachandran and T Yamagata (1999) "A dipole mode in the tropical Indian Ocean" *Nature* 401: 360-363

Saunders, D A, A J M Hopkins and R A How (Eds) (1990) *Australian Ecosystems: 200 Years of Utilization, Degradation and Reconstruction* Proceedings of the Ecological Society of Australia 16

Schneider, A and H Ingram (1988) "Systematically Pinching Ideas: A Comparative Approach to Policy Design" *Journal of Public Policy* 8(1): 61-80

Schram, S F (1995) *Words of Welfare. The Poverty of Social Science and the Social Science of Poverty* Minneapolis, University of Minnesota Press

Schultz, J (1998) *Reviving the Fourth Estate: Democracy, Accountability and the Media* Melbourne, Cambridge University Press

Senate Standing Committee on Rural and Regional Affairs (1992) *A national drought policy—appropriate government responses to the recommendations of the Drought Policy Review Task Force: Final report* Canberra, The Parliament of the Commonwealth of Australia

Shakespeare, W (1996) *King Lear* London, Penguin Books

Shanteau, J (1992) "Decision making under risk: Applications to insurance purchasing" *Advances in Consumer Research* 19: 177-181

Shattuck, J (1996) "Human rights and humanitarian crisis: Policy making and the media" in R I Rotberg and T G Weiss (Eds), *From Massacres to Genocide* Washington, Brookings Institution

Shepard, M (1999) *The Simpson Desert: Natural History and Human Endeavour* Adelaide, Corkwood Press

Sigurdson, D and R Sin (1994) "An Aggregate Analysis of Canadian Crop Insurance Policy" in D L Heuth and W H Furtan (Eds), *Economics of Agricultural Crop Insurance: Theory and Evidence* Boston/Dordrecht/London, Kluwer Academic Publishers: 45-72

Simmons, P (1993) "Recent Developments in Commonwealth Drought Policy" *Review of Marketing and Agricultural Economics* 61(3): 443-454

Simon, H A (1953) *Administrative behaviour: A study of decision-making processes in administrative organisation* New York, The Macmillan Company

Simpson, J (1997) "Perception of Meteorology in some Aboriginal Languages" in E Webb (Ed.) *Windows on Meteorology: Australian Perspective* Melbourne, CSIRO Publishing: 20-28

Skees, J (1999) *Agriculture Insurance in a Transition Economy* OECD Meeting on Agricultural Finance and Credit Infrastructure in Transition Economies Moscow

Skees, J (2001) "The Bad Harvest" *Regulation* (Spring): 16-21

Smith, D I, M F Hutchinson and R J McArthur (1992) *Climatic and Agricultural Drought: Payments and Policy* Report to the Rural Industries Research and Development Corporation, Centre for Resource and Environmental Studies, The Australian National University

Smith, M, E Williams and R Wasson (1991) "The Archaeology of the JSN Site: Some Implications for the Dynamics of Human Occupation in the Strzelecki Desert During the Late Pleistocene" *Records of the South Australia Museum* 25(2): 175-192

Snedden, t H B, QC MP (1971) *Commonwealth Payments to or for the States 1971—72* Parliamentary Paper No 54 Canberra, Commonwealth Government Printing Service, 17 August 1971

Sood, R, G Stockdale and E M Rogers (1987) "How the news media operate in natural disasters" *Journal of Communication* 37(3): 27-41

Spencer, B (Ed.) (1994) *Report on the work of the Horn Scientific Expedition to central Australia. Part I. Introduction, narrative, summary of results, supplement to zoological report, map* Bundaberg, Queensland, Facsimile—Corkwood Press

Stafford Smith, D M (1994a) *Sustainable Production Systems and Natural Resource Management in the Rangelands* Proceedings ABARE Outlook Conference Canberra, ABARE

Stafford Smith, D M, J F Clewett, A M Moore, G M McKeon and R Clark (1997) *DroughtPlan. Full Project Report.* DroughtPlan Working Paper No 10 Canberra, CSIRO Alice Springs/LWRRDC Occasional Paper Series

Stafford Smith, D M and S R Morton (1990) "A framework for the ecology of arid Australia" *Journal of Arid Environments* 18: 225-278

Stafford Smith, M (1994b) *A regional framework for managing the variability of production in the rangelands of Australia* Alice Springs, CSIRO/RIRDC

Stafford Smith, M (2003a) "Linking environments, decision-making and policy in handling climate variability" in L C Botterill and M Fisher (Eds), *Beyond Drought: People, Policy and Perspectives* Melbourne, CSIRO Publishing: 131-151

Stafford Smith, M (2003b) "Living in the Australian environment" in L C Botterill and M Fisher (Eds), *Beyond Drought: People, Policy and Perspectives* Melbourne, CSIRO Publishing: 9-20

Stafford Smith, M, R Buxton, G McKeon and A Ash (2000) "Seasonal climate forecasting and the management of rangelands: do production benefits translate into enterprise profits?" in G L Hammer, N Nicholls and C Mitchell (Eds), *Applications of seasonal climate forecasting in agricultural and natural ecosystems* Dordrecht, Kluwer Academic Publishers: 271-289

Stafford Smith, M and G M McKeon (1998) "Assessing the historical frequency of drought events on grazing properties in Australian rangelands" *Agricultural Systems* 57: 271-299

Stafford Smith, M, S R Morton and A J Ash (2000) "Towards sustainable pastoralism in Australia's rangelands" *Australian Journal of Environmental Management* 7(4): 190-203

Stafford Smith, M and J F Reynolds (2002) "Desertification: A new paradigm for an old problem" in J F Reynolds and D M Stafford Smith (Eds), *Dahlem Workshop Report 88* Berlin, Dahlem University Press: 403-425

States and Northern Territory Grants (Rural Adjustment) Act 1988 Act No. 112 of 1988

Stehlik, D (1999) "Developing a Rural Consciousness: Aged Care, Rural Health & Rural Well-Being for Sustainable Australian Communities" in W Ramp, J Kulig, I Townshend and V McGowan (Eds), *Health in Rural Settings: Contexts for Action* Lethbridge: Canada, University of Lethbridge: 143-162

Stehlik, D, H Bulis, I Gray and G Lawrence (1996) *Communities in crisis? Towards a concept of quality of life in Australian rural communities who have experienced the drought of the early 1990s* Paper presented to the International Quality of Life Conference Price George, British Columbia, University of Northern British Columbia, Canada

Stehlik, D, I Gray and G Lawrence (1999) *Drought in the 1990s. Australian Farm Families' Experiences* Canberra, Rural Industries Research and Development Corporation 99/14, UCQ-5A

Stehlik, D and G Lawrence (1996) "Rural Policy. A Contradiction in Terms? The Example of Aged Care Policy in Australia" *Journal of the Australian Studies Institute* 3(1): 91-108

Stehlik, D and G Lawrence (1999) *Key Issues and Strategic Directions Relating to Rural and Remote Communities* Discussion Paper prepared for Queensland Department of Premier and Cabinet, July

Stehlik, D, G Lawrence and I Gray (2000) "Gender and Drought: Experiences of Australian Women in the Drought of the 1990s" *Disasters* 24(1): 38-53

Stephens, D J (1998) "Objective criteria for estimating the severity of drought in the wheat cropping areas of Australia" *Agricultural Systems* 57: 333-350

Stone, R and G de Hoedt (2000) "The development and delivery of seasonal climate forecast capabilities in Australia" in G L Hammer, N Nicholls and C Mitchell (Eds), *Applications of seasonal climate forecasting in agricultural and natural ecosystems* Dordrecht, Kluwer Academic Publishers: 67-76

Strang, V (2001) "Poisoning the rainbow" in A Rumsey and J Weiner (Eds), *Mining and Indigenous Lifeworlds in Australia and Papua New Guinea* Adelaide, Crawford House Publishing: 208-225

Sturman, A P and N Tapper (1996) *The Weather and Climate of Australia and New Zealand* Melbourne, Oxford University Press

Synapse Consulting (Aust) Pty Ltd (1992) *Report of the Review of the Rural Adjustment Scheme* Brisbane

Tadesse, T (2000) "Drought and its Predictability in Ethiopia" in D A Wilhite (Ed.) *Drought: A Global Assessment* London, Routledge Volume 1: 135-142

Terkildsen, N, F Schnell and C Ling (1998) "Interest groups, the media, and policy debate formation. An analysis of message structure, rhetoric and source cues" *Political Communication* 15(1): 46-61

Thompson, D and R Powell (1998) "Exceptional Circumstances Provisions in Australia—Is There too Much Emphasis on Drought" *Agricultural Systems* 57(3): 469-488

Tindale, N (1974) *Aboriginal Tribes of Australia: Their Terrain, Environmental Controls, Distribution, Limits, and Proper Names* Canberra, Australian National University Press

Treasury (2003) *Mid Year Economic and Fiscal Outlook* Canberra

Treasury (2004) *Tax Expenditures Statement* Canberra

Trewin, R, D Peterson, R Chambers and L Moon (1992) *Managing farm risk* Presentation to National Agricultural and Resources Outlook Conference Canberra, Australian Bureau of Agricultural and Resource Economics

Truss, W, the Hon MP (2002) *Commonwealth to push States/Territories to put drought-stricken farmers first* Media Release by the Federal Minister for Agriculture, Fisheries and Forestry, 2 October 2002

Truss, W, the Hon MP (2003a) *Sustaining agriculture—the drought and beyond* Media Release by Federal Minister for Agriculture, Fisheries and Forestry AFFA03/124WT, 13 May 2003

Truss, W, the Hon MP (2003b) *Primary Industries Ministerial Council Meeting Communique* Media Release by Federal Minister for Agriculture, Fisheries and Forestry PIMC 4/03 Canberra, 3 October 2003

Tuchman, G (1997) "Making news by doing work. Routinising the unexpected" in D Berkowitz (Ed.) *Social Meanings of News* Thousand Oaks, Sage

Ubergang, J W (2002) *The Crooble Plan: To reduce the vulnerability of Australia to weather and climate extreme disasters* North West Magazine December 2: 6

UNCCD (1999) *United Nations Convention to Combat Desertification (text with annexes)* Bonn, Germany

UNDP/UNSO (2000) *Report on the Status of Drought Preparedness and Mitigation in Sub-Saharan Africa* New York, UN Development Program, Office to Combat Desertification and Drought

Van Manen, M (1990) *Researching Lived Experience. Human Science for an Action Sensitive Pedagogy* New York, State University of New York Press

Vincent, D, A Watson and L Barton (1975) "Poverty among farmers in three districts in Victoria" in R F Henderson (Ed.) *Financial aspects of rural poverty* Canberra, AGPS

Vincent, R, B Crow and D Davis (1997) "When technology fails. The drama of airline crashes in network television news" in D Berkowitz (Ed.) *Social Meanings of News* Thousand Oaks, Sage

Wahlquist, Å (2003) "Media representations and public perceptions of drought" in L C Botterill and M Fisher (Eds), *Beyond Drought: People, Policy and Perspectives* Melbourne, CSIRO Publishing: 67-86

Walcott, J J and A J Clark (2001) *Risk assessment and expert opinion in implementing policy—Exceptional Circumstances* Proceedings of the 10th Australian Agronomy Conference Hobart http://www.regional.org.au/au/asa/2001/3/a/walcott.htm#TopOfPage

Walker, K, J Puckridge and S Blanch (1997) "Irrigation development on Cooper Creek, central Australia— prospects for a regulated economy in a boom-and-bust ecology" *Aquatic Conservation: Marine and Freshwater Ecosystems* 7: 63-73

Walker, K, F Sheldon and J Puckridge (1995) "A Perspective on Dryland River Ecosystems" *Regulated Rivers: Research and Management* 11: 85-104

Walsh, P (1994) "Cassandra" *Australian Financial Review* 18 July 1994

Walsh, P, Senator (1989) *Question without Notice: Natural Disaster Assistance: Queensland* Senate Hansard Canberra 2 March 1989

Ward, I (1995) *Politics of the Media* South Melbourne, Macmillan

Waterford, J (2002) *Dealing differently with drought* Australian Policy Online http://www.apo.org.au/webboard/items/00142.shtml

Watts, M (2003) *Reporting Unemployment* George Munster Forum Newcastle http://journalism.uts.edu.au/acij/forums/2003forum1a.html

Webb, P, J von Braun and Y Yohannes (1992) *Famine in Ethiopia: Policy implications of coping failure at national and household levels* International Food Policy Research Institute Research Report No 92 Washington DC

Weinstein, N D and M Nicolich (1993) "Correct and incorrect interpretations of correlations between risk perceptions and risk behaviours" *Health Psychology* 12(3): 235-245

Wenger, D and B Friedman (1986) "Local and national media coverage of disaster: A content analysis of the print media's treatment of disaster myths" *International Journal of Mass Emergencies and Disasters* 4(3): 27-50

West, B and P Smith (1996) "Drought, discourse and Durkheim: A research note" *Australian and New Zealand Journal of Sociology* 32(1): 93-102

Whetton, P H, J J Katzfey, K J Hennessy, X Wu, J L McGregor and K C Nguyen (2001) "Developing scenarios for climate change for Southeastern Australia: An example using regional climate model output" *Climate Research* 16: 181-201

White, B J (2000a) "The importance of climate variability and seasonal forecasting to the Australian economy" in G L Hammer, N Nicholls and C Mitchell (Eds), *Applications of seasonal climate forecasting in agricultural and natural ecosystems—the Australian experience* The Netherlands, Kluwer Academic: 1-22

White, D H (1998) "Editorial Introduction" *Agricultural Systems* 57(3): 227-229

White, D H and V M Bordas (Eds) (1996) *Indicators of Drought Exceptional Circumstances. Proceedings of a Workshop held in Canberra on 1 October 1996* Canberra, Bureau of Resource Sciences

White, D, D Collins and M Howden (1993) "Drought in Australia: Prediction, monitoring, management and policy" in D A Wilhite (Ed.) *Drought Assessment, Management, and Planning: Theory and Case Studies* Boston, Kluwer: 213-236

White, D H, S M Howden, J J Walcott and R M Cannon (1995) *Estimating the extent and variability of drought* International Congress on Modelling and Simulation Proceedings. MODSIM. Vol. 2: Air Pollution and Climate. The University of Newcastle

White, D H, S M Howden, J J Walcott and R M Cannon (1998) "A framework for estimating the extent and severity of drought, based on a grazing system in south-eastern Australia" *Agricultural Systems* 57: 259-270

White, D H and L Karssies (1997) *Australia's National Drought Policy: Aims, analyses and implementation* IXth World Water Congress Montreal, Canada

White, D H, G Tupper and H Mavi (1999a) *Climate variability and drought research in relation to Australian agriculture: Research compendium* LWRRDC Occasional Paper CV01/99 Canberra, Land and Water Resources Research & Development Corporation http://www.acslink.aone.net.au/asit/Compend98.htm

White, D H, G Tupper and H S Mavi (1999b) *Agricultural climate research and services in Australia* Land and Water Resources Research and Development Corporation LWRRDC Occasional Paper CV02/99 Canberra

White, M (2000b) *Running Down: Water in a Changing Land* Sydney, Kangaroo Press

White, W B (2000c) "Influence of the Antarctic Circumpolar Wave on Australian precipitation from 1958 to 1997" *Journal of Climate* 13: 2125-2141

Wilhite, D A (1991) "Drought Planning: A Process for State Government" *Water Resources Bulletin* 27(1): 29-38

Wilhite, D A (1993) "The Enigma of Drought" in D A Wilhite (Ed.) *Drought Assessment, Management, and Planning: Theory and Case Studies* Boston, Kluwer Academic Publishers: 3-15

Wilhite, D A (1997) "State Actions to Mitigate Drought: Lessons Learned" *Journal of the American Water Resources Association* 33(5): 961-68

Wilhite, D A (2000a) "Drought as a Natural Hazard: Concepts and Definitions" in D A Wilhite (Ed.) *Drought: A Global Assessment* London, Routledge Volume 1: 3-18

Wilhite, D A (Ed.) (2000b) *Drought: A Global Assessment* London, Routledge Publishers

Wilhite, D A (2001) "Moving Beyond Crisis Management" *Forum for Applied Research and Public Policy* 16(1): 20-28

Wilhite, D A and M H Glantz (1985) "Understanding the Drought Phenomenon: The Role of Definitions" *Water International* 10: 111-120

Wilhite, D A and M H Glantz (1987) "Understanding the drought phenomenon: The role of definitions" in D A Wilhite, W E Easterling and D A Wood (Eds), *Planning for Drought: Toward a Reduction of Societal Vulnerability* Boulder, Colorado, Westview Press: 11-30

Wilhite, D A, M J Hayes, C Knutson and K H Smith (2000) "Planning for Drought: Moving from Crisis to Risk Management" *Journal of American Water Resources Association* 36: 697-710

Wilhite, D A, M V K Sivakumar and D A Wood (2000) "Improving Drought Early Warning Systems in the Context of Drought Preparedness and Mitigation, Summary of Breakout Sessions" in D A Wilhite, M V K Sivakumar and D A Wood (Eds), *Early Warning Systems for Drought Preparedness and Drought Management, Proceedings of an Expert Group Meeting, Lisbon, Portugal, 5-7 September* Geneva, Switzerland, World Meteorological Organization

Wilhite, D A and O Vanyarkho (2000) "Drought: Pervasive Impacts of a Creeping Phenomenon" in D A Wilhite (Ed.) *Drought: A Global Assessment* London, Routledge Publishers Volume 1: 245-255

Wilkins, L (1986) "Media coverage of the Bhopal Disaster: A cultural myth in the making" *International Journal of Mass Emergencies and Disasters* 4(1): 7-33

Wilson, E (2002) *The Future of Life* New York, Alfred A. Knopf

Winters, L A (1990) "The so-called 'non-economic' objectives of agricultural support" *OECD Economic Studies* 13: 237-266

Yunupingu, B (1991) "A plan for ganma research" in J Henry and R McTaggart (Eds), *Aboriginal Pedagogy: Aboriginal Teachers Speak Out* Geelong, Deakin University Press: 98-106

Zhang, P G (1998) *Exotic Options* (Second edition) Singapore, World Scientific Publishing Co

INDEX

Advances in Natural and Technological Hazards Research

Series Editor: Prof. Dr. Mohammed I. El-Sabh, *Département d'Océanographie, Université du Québec à Rimouski, 310 Allée des Ursulines, Rimouski, Québec, Canada G5L 3A1*

Publications

1. S. Tinti (ed.): *Tsunamis in the World.* Fifteenth International Tsunami Symposium (1991). 1993   ISBN 0-7923-2316-5

2. J. Nemec, J.M. Nigg and F. Siccardi (eds.): *Prediction and Perception of Natural Hazards.* Symposium Perugia, Italy (1990). 1993
         ISBN 0-7923-2355-6

3. M.I. El-Sabh, T.S. Murty, S. Venkatesh, F. Siccardi and K. Andah (eds.): *Recent Studies in Geophysical Hazards.* 1994   ISBN 0-7923-2972-4

4. Y. Tsuchiya and N. Shuto (eds.): *Tsunami: Progress in Prediction, Disaster Prevention and Warning.* 1995   ISBN 0-7923-3483-3

5. A. Carrara and F. Guzzetti (eds.): *Geographical Information Systems in Assessing Natural Hazards.* 1995   ISBN 0-7923-3502-3

6. V. Schenk (ed.): *Earthquake Hazard and Risk.* 1996   ISBN 0-7923-4008-6

7. M.I. El-Sabh, S. Venkatesh, H. Denis and T.S. Murty (eds.): *Land-based and Marine Hazards.* Scientific and Management Issues. 1996
         ISBN 0-7923-4064-7

8. J.M. Gutteling and O. Wiegman: *Exploring Risk Communication.* 1996
         ISBN 0-7923-4065-5

9. G. Hebenstreit (ed.): *Perspectives on Tsunami Hazard Reduction.* Observations, Theory and Planning. 1997   ISBN 0-7923-4811-7

10. C. Emdad Haque: *Hazards in a Fickle Environment: Bangladesh.* 1998
         ISBN 0-7923-4869-9

11. F. Wenzel, D. Lungu and O. Novak (eds.): *Vrancea Earthquakes: Tectonics, Hazard and Risk Mitigation.* 1999   ISBN 0-7923-5283-1

12. S. Balassanian, A. Cisternas and M. Melkumyan (eds.): *Earthquake Hazard and Seismic Risk Reduction.* 2000   ISBN 0-7923-6390-6

13. S.L. Soloviev, O.N. Solovieva, C.N. Go, K.S. Sim and N.A. Shchetnikov: *Tsunamis in the Mediterranean Sea 2000 B.C. – 2000 A.D.* 2000
         ISBN 0-7923-6548-8

14. J.V. Vogt and F. Somma (eds.): *Drought and Drought Mitigation in Europe.* 2000   ISBN 0-7923-6589-5

Advances in Natural and Technological Hazards Research

15. M. Oya: *Applied Geomorphology for Mitigation of Natural Hazards.* 2001
ISBN 0-7923-6719-7

16. E. Coles, D. Smith and S. Tombs (eds.): *Risk Management and Society.* 2001
ISBN 0-7923-6899-1

17. T. Glade, P. Albini and F. Francés: *The Use of Historical Data in Natural Hazard Assessments.* 2001 ISBN 0-7923-7154-2

18. G.T. Hebenstreit (ed.): *Tsunami Research at the End of a Critical Decade.* 2001 ISBN 1-4020-0203-3

19. J. Locat and J. Mienert (eds.): *Submarine Mass Movements and Their Consequences.* 1st International Symposium. 2003 ISBN 1-4020-1244-6

20. K.F. O'Loughlin and J.F. Lander: *Caribbean Tsunamis.* A 500-Year History from 1498-1998. 2003 ISBN 1-4020-1717-0

21. J.P. Stoltman, J. Lidstone and L.M. DeChano (eds.): *International Perspectives on Natural Disasters: Occurrence, Mitigation, and Consequences.* 2004
ISBN 1-4020-2850-4

22. L.C. Botterill and D.A. Wilhite (eds.): *From Disaster Response to Risk Management.* Australia's National Drought Policy. 2005 ISBN 1-4020-3123-8